"Tim Unwin's new book digs into some of the deep psychological reasons why our digital interventions so often fail to live up to the promise of being 'for all'. He offers a sobering, even dystopian view of where things stand today—yet, at the same time, he inspires with his optimism that we still have the power to shape such technologies for real and lasting good."

Ralph Hertwig, *Professor and Director of the Center for Adaptive Rationality, Max Planck Institute for Human Development, Berlin*

"Tim's diverse experiences have given him unique insights into what needs to be done so that the underprivileged and most marginalized population in Africa can really benefit from the use of digital tech. Used appropriately, digital tech can bring access to education, healthcare, financial services, and employment opportunities. It is an essential tool for social inclusion with the potential to reduce inequalities and foster sustainable development. This challenging book combines his practical experiences with rigorous scholarship to create a highly readable manifesto that cuts through the rhetoric and tells us all what we really need to do."

Helena Fernandes, *Chairperson of the Board of Directors, Mozambique Communications Regulatory Authority (INCM)*

"This new book raises alarms about really important matters. The gap in digital access and use is getting bigger, and the explicit intent for equality and equity in global initiatives is so rarely there. This powerful book resonates with many of my own concerns."

Haif Bannayan, *Global Director Business Development, The International Baccalaureate*

"I share with Tim a belief that we need to continue to do more to support the poorest and most marginalised individuals and communities in the world. What I do not share with him is his deep understanding of how digital tech can contribute to this effort, but equally how it can have profoundly harmful effects and deepen divisions and inequalities. As a result of reading this book, I am at least a little closer to that understanding, and hope that policy makers and those advising them will take careful note of its wise words and recommendations."

Sir Myles Wickstead, *Visiting Professor, International Relations, and former senior civil servant and Ambassador*

"Tim has articulated what many digital development advocates have long observed and privately lamented but seldom voiced publicly. In a world shaped by geopolitical tension, self-interest, and corporate exploitation, the gains of the technological revolution have reached only a privileged few. His book urges a return to compassion and reframes equity as an achievable goal. He critiques the corporate and political ecosystems, and makes an inspiring call for digital emancipation, laying out a clear roadmap for transformation, closing digital divides and unlocking opportunity for communities everywhere."

Bernadette Lewis, *Secretary General, Commonwealth Telecommunications Organisation*

"Digital transformation is having a major influence on the economy and the future of humanity, but many people in developing countries are still seeing only limited benefits. This important new book challenges us all to think again about how we can ensure that everyone benefits from the use of digital tech. Not only does it provide a sound critique, it also offers a clear manifesto for what we urgently need to do to rectify this."

Torbjörn Fredriksson, *Head of E-commerce and Digital Economy, UNCTAD*

"Tim has crossed many boundaries between academia, government service, international organisations, civil society and the private sector. This powerful and passionate book draws on his very extensive research practice to present a convincing case for why digital tech has not yet sufficiently been used to serve the world's poorest and most marginalised. It also goes beyond this, and tells us clearly what we need to do to make the world fairer and less divided."

Chaesub Lee, *Chairman of Passwordless Alliance and Principal Advisor of Welchmann Keen; formerly Director of ITU Telecommunication Standardization Bureau*

Digital Inclusion in an Unequal World

Why do so many attempts to use digital tech to help the world's poor ultimately end in failure? *Digital Inclusion in an Unequal World* provides a detailed critique of previous theory and practice, while also proposing practical and realistic suggestions for good practice.

This book combines longer thematic chapters by Tim Unwin, one of the world's leading thinkers on ICT for development, with shorter vignettes from other experts across a range of different practical, intellectual, and geographical backgrounds. The book argues that the global geo-politico-economic agendas associated with the use of digital tech in development in late-capitalism raise pressing issues around instrumentalism, individualism, and empowerment. This context drives short-termism and an innovation fetish around current hot topics, such as EdTech, Blockchain, and Artificial Intelligence. The world's poorest and most marginalised people are failing to benefit from the use of digital technology, whilst the world's Digital Barons continue to accrue profits. Unwin emphasises the importance of crafting a responsibilities agenda that will shift the dial, enabling practitioners to avoid common pitfalls and transform good intent into good practices.

This book is a highly readable guide for the global community of development practitioners, government officials, and civil society organisations involved in delivering digital tech initiatives. Students from the fields of international development, computer science, electronic engineering, geography, and economics will also benefit from its expert insights.

Tim Unwin is Emeritus Professor of Geography at Royal Holloway, University of London, UK, where his research-practice over the last 25 years has focused especially on the use of digital technologies by the world's poorest and most marginalised people. He is the Founder of the ICT4D Collective (2004–present), was Chairholder of the UNESCO Chair in ICT4D (2007–2023), and served as Secretary General of the Commonwealth Telecommunications Organisation (2011–2015).

Rethinking Development

Rethinking Development offers accessible and thought-provoking overviews of contemporary topics in international development and aid. Providing original empirical and analytical insights, the books in this series push thinking in new directions by challenging current conceptualizations and developing new ones.

This is a dynamic and inspiring series for all those engaged with today's debates surrounding development issues, whether they be students, scholars, policy makers and practitioners internationally. These interdisciplinary books provide an invaluable resource for discussion in advanced undergraduate and postgraduate courses in development studies as well as in anthropology, economics, politics, geography, media studies and sociology.

Reformers in International Development
Five Remarkable Lives
David M. de Ferranti

Young People in the Global South
Voice, Agency and Citizenship
Edited by Kate Pincock, Nicola Jones, Lorraine Van Blerk and Nyaradzayi Gumbonzvanda

Community-Led Development in Practice
We Power our Own Change
Edited by Elene Cloete and Gunjan Veda

Post-Development from the Global South
Radical Alternatives or Ambivalent Engagements?
Edited by Sally Matthews and Alba Castellsagué

Digital Inclusion in an Unequal World
An Emancipatory Manifesto
Tim Unwin

For more information about this series, please visit: www.routledge.com/Rethinking-Development/book-series/RDVPT

Digital Inclusion in an Unequal World

An Emancipatory Manifesto

Tim Unwin

Routledge
Taylor & Francis Group
LONDON AND NEW YORK

Designed cover image: Tim Unwin

First published 2026
by Routledge
4 Park Square, Milton Park, Abingdon, Oxon OX14 4RN

and by Routledge
605 Third Avenue, New York, NY 10158

Routledge is an imprint of the Taylor & Francis Group, an informa business

British Library Cataloguing-in-Publication Data
A catalogue record for this book is available from the British Library

Library of Congress Cataloging-in-Publication Data
A catalog record for this book has been requested

ISBN: 978-1-032-98306-6 (hbk)
ISBN: 978-1-032-98304-2 (pbk)
ISBN: 978-1-003-59800-8 (ebk)

DOI: 10.4324/9781003598008

Typeset in Times New Roman
by Apex CoVantage, LLC

Contents

Figures

Contributors

Shabnam Aggarwal is the CTO of Rising Academies, an AI strategist, the author of *Freedom to Fail*, and has spent 15 years building and breaking education technology across Asia and Africa. Her start-ups, failures and comebacks fuel her mission to design human-centred, low-cost tools that live for decades in classrooms. She believes failure is a feature, not a bug.

Ken Banks is a social innovator, author, and technologist known for developing tools that empower grassroots change. He founded kiwanja.net and created FrontlineSMS, helping communities worldwide. His work bridges technology, development and impact, earning global recognition for supporting social entrepreneurs, local leaders, and underrepresented voices through practical, human-centred solutions.

Tendani Mulanga Chimboza is a passionate and driven professional committed to making a difference through inclusive leadership, academic excellence, and impactful research. She advocates for representation and social justice, using technology to empower communities. Her work bridges academia (University of Cape Town, South Africa), corporate insight, and public engagement to drive ethical digital transformation across Africa.

Marine Al Dahdah is a sociologist and a CNRS Fellow at the Center for the Study of Social Movements (EHESS-Paris) and the French Institute of Pondicherry in India. Her research focuses on digital policies in Asia and Africa and more particularly on digital health in India, Ghana, and Kenya.

Pari Esfandiari works at the intersection of technology, geopolitics, and governance. She is President and Co-Founder of the Global TechnoPolitics Forum, advises international bodies, serves on ICANN's At-Large Advisory Committee, was a Senior Fellow at the Atlantic Council, and helps leaders navigate digital transformation and global systems beyond borders.

Domenico Fiormonte is Associate Professor at the University of Roma Tre, Italy, and Research Associate at the University of Dar es Salaam, Tanzania, where he teaches courses on Sociology of Communication, Geopolitics of Knowledge, and Digital Humanities.

Mei Lin Fung is the Co-Founder of the People-Centered Internet with Vint Cerf, and is a tech pioneer in CRM and the future of health. She leads global efforts in digital public infrastructure, focused on bridging the gap so global finance can reach MSMEs everywhere.

Akber Gardezi is an academic and entrepreneur. As founding head of Data and AI at Mobiz and a tenured Associate Professor at COMSATS University Islamabad, Pakistan, he champions inclusive technology across Islamic nations. His work, grounded in research and innovation, reflects scholarly rigour, entrepreneurial vision, and community empowerment.

G. 'Hari' Harindranath is Professor of Information Systems at Royal Holloway, University of London, UK. He explores how digital tech shapes lives, work, and society. Hari is passionate about impact, collaborates with communities to address digital inequalities, and sometimes even gets policy-makers to listen. He thinks research should do good, not just look good.

Judith Hellerstein is a globally recognised expert in ICT policy, digital transformation, cyber, and digital development. She has over 30 years of experience in crafting policies that enable digital economies, build digital infrastructure, e-commerce frameworks, and digital trade. Judith is a passionate advocate for helping to bring connectivity and empower communities.

Emily Hickson is an international climate politics expert, having advised businesses and civil society on the UN Climate Negotiations (COPs) for the last decade. She is also a local government councillor in London, UK.

David Hollow is Co-Founder of Jigsaw Education and the Research Director of EdTech Hub in the UK. He enjoys building practical evidence to improve education outcomes. He provides strategic advice to governments and donors and serves on various boards. Many years ago, David was privileged to have Tim Unwin as a wonderful PhD supervisor!

Nick Hughes OBE is an innovator and practitioner, passionate about building digital ventures that help solve real-world problems. Nick's work spans Africa (e.g. M-PESA and M-KOPA) and Asia (e.g. bKash). 4R Digital Ltd is his latest venture, where he and his team apply digital technology to bring radical transparency and more equitable flow of funds into B2B value chains.

Poncelet Ileleji is a computer scientist and ICT expert in The Gambia dedicated to leveraging technology for sustainable development. He is the Lead and CEO of Jokkolabs Banjul, a board member of www.apc.org, a former President of the Information Technology Association of The Gambia, and an active voice in Internet governance, digital learning, and innovation across Africa.

Emmanuel C. Lallana has 25 years' experience in Digital Development across all sectors, conducting research, policy development, and training for government officials from developing Asian countries on digital policy issues, such as ICT for Education, Digital Transformation and public interest AI. He served on the national Commission on Information and Communications Technology 2004–2007 in the Philippines.

Janet Longmore is Founder and CEO of Digital Opportunity Trust (DOT), and is a lifelong social entrepreneur who champions the leadership of young women and men in global development. She is the recipient of many Canadian and international honours, and demonstrates the power of technology to unlock the creativity of youth as leaders of change.

Yuliya Morenets is a barrister at law, a digital governance expert, and the Founder of the Youth IGF. A recognised leader in global digital cooperation, she advises governments on cybersecurity and cybercrime. With over 15 years of international experience, she is also a media advocate and champion of digital inclusion in the Global South.

Nnenna Nwakanma, also known as 'Nnenna from the Internet', is a voice for systemic policy changes for development through open, rights-based, and human-centred technology. She is a champion for empowering Internet and information access, freedom of expression, open data, open government, and the open web. She was proclaimed one of the 100 Most Influential Africans of 2021.

Dorothy Okello is Dean, School of Engineering, and Associate Professor, Department of Computer and Communications Engineering at Makerere University, Uganda. She is a leader with diverse experience in teaching, researching, and conducting projects and contributing to inclusive policies, including serving on boards in education, telecommunications/ICT, banking, and innovation.

Alejandro Pisanty is a Professor, and former Head of Academic Computing Services, and Coordinator for Open and Distance Education at the National University of Mexico. He has been a member of the Board of Directors of ICANN and of the Board of Trustees of ISOC. In 2021, he was inducted to the Internet Hall of Fame.

Jamie Proctor has worked in tech start-ups, across UK Government digital teams, and established a school construction NGO in Malawi. Jamie's background cuts across research, education, and technology. He wrote this contribution whilst seconded to EdTech Hub from FCDO, and maintains keen research interests in the development of climate-resilient and sustainable school infrastructure.

Nimmi Rangaswamy is Professor at the Kohli Centre on Intelligent Systems, IIIT, Hyderabad, India. She teaches on the society-technology interface, bringing an anthropological perspective to explore the impacts of AI. Her research in Human-Computer Interaction foregrounds local practices of technology diffusion and how communities engage with digital systems – through sharing, learning, training, and adaptation.

Benita Rowe designs and advises on digital strategies with a focus on service delivery in humanitarian and conflict-affected settings. Over nearly two decades, she has worked across crisis preparedness and response, education and health, supporting system resilience across more than 17 countries, in collaboration with UN agencies, international organisations, private-sector and industry actors, local organisations, and foundations.

Fernanda Scur is a multi-stakeholder partnership strategist, aimed at purposeful engagement and effective collaborative action that drives systemic/structural change for positive social and environmental impact. She holds a PhD in Systems Thinking for Active Collaboration and has experience in interaction design, project development, concept and process design, complex network analysis and user experience.

Paul Spiesberger graduated as a computer scientist from TU Wien, Austria, and is a passionate programmer. He is the operative head of ROTA at INSO and works with SAI, TU Graz. Since 2014, he has served as the Chair of ICT4D.at, and is a member of the ICT4D Collective.

Revi Sterling started out elevating women's voices through technology. This led to positions and programmes helping millions of women use technology to achieve their goals, as well as 20+ years of frustration with donors, governments, peer institutions and rural buses. She will never stop talking about social norms, opportunities, and inequities.

Mohamed Sharil Tarmizi trained and practised as a lawyer, became a policy-maker, went into corporate finance and then moved on to become a Telco regulator and a trade negotiator. Thereafter, he became involved in corporate restructuring, automotive, infrastructure and aviation, whilst dabbling in a spot of angel investing, startup mentoring, media and movie making.

Juliette Unwin is a Senior Lecturer in Statistical Science at the University of Bristol, UK, where her research focuses on developing methodologies related to infectious disease transmission to improve global public health. She is passionate about co-creating research with partners in areas where the global burden of disease is highest, whilst sharing her technical knowledge when appropriate.

Ugo Vallauri is Co-Founder/Co-Director of The Restart Project, a charity which promotes repair to change our throwaway economy. He is also a Co-Founder of Right to Repair Europe, a coalition advocating for ambitious right to repair legislation. Ugo is a Fellow of the Ashoka Foundation in Italy, with 20 years of experience in the not-for-profit sector.

Lousewies van der Laan is Director of Transparency International Netherlands and advises the Dutch Authority on Online Terrorist and CSAM. She is a former Member of Parliament and served on the ICANN board. She is Ambassador for the Netherlands Internet Governance Forum and recipient of the ISOC NL Lifetime Achievement Award.

Tom Wambeke is a UN Senior Executive and Chief of Learning Innovation at ITCILO, Italy. With global experience in learning innovation, digital transformation, and foresight, he leads impactful programmes on inclusive capacity development. A speaker, strategist, and facilitator, he collaborates worldwide to foster sustainable, tech-enabled solutions for digital inclusion.

Preface

This book has three main purposes. First, it sheds *light* on why the poorest and most marginalised are still being left behind by the latest advances in digital technologies, despite global rhetoric otherwise. Second, it seeks to *understand* why many of those involved in using digital technologies for 'development' resolutely seem to fail to learn from previous practice and research, and therefore continue to make similar mistakes to those we have made in the past. Third, it advances an *emancipatory* agenda based on responsibilities rather than rights, that has the power to change existing priorities and practices. The words 'light', 'understanding', and 'emancipation' situate it firmly within a tradition of Critical Theory that draws much on the work of Jürgen Habermas (1974; 1978): 'light' reflects an Enlightenment belief that by 'understanding' what is happening in the world, we may be in a stronger place to change it for the better; and 'emancipation' implies that by so doing, we can free ourselves and others from the illusionary shackles of domination and control that have been used by those in power to enslave and surveil us. The focus on responsibilities rather than rights also reflects my debt to Onora O'Neill (1996; 2016) whose work on obligations and justice across boundaries has served as a constant source of inspiration, although I persist in preferring to address the notion of 'responsibilities' rather than adhering to her focus on 'obligations'.

The book is the final work in an unintended trilogy, following my edited *ICT4D: Information and Communication Technology for Development* (Unwin, 2009) and my *Reclaiming Information and Communication Technologies for Development* (Unwin, 2017a). The first of these was the most optimistic, and sought to outline a framework for Information and Communication Technology for Development (ICT4D) that was fundamentally situated within the broad field of 'development', rather than in computer science or information systems (for which, see Heeks, 1995; Toyama, 2015; Walsham, 2012; 2017). It also drew on the experiences of many of its authors in trying to implement practical interventions on the ground and share the lessons they had learnt from so doing. However, it did not shy away from asking the question 'What needs to be in place if the poorest and most vulnerable people and communities are to be able to take advantage of the potential benefits that ICTs can provide?' (Unwin, 2009, p. 3; see also Milward-Oliver, 2005). The second of these, *Reclaiming Information and Communication Technologies for*

Development was written almost a decade later in 2017, partly in frustration that little seemed to have changed over the intervening years. It suggested that one of the main reasons why so few benefits were being experienced by the economically poor was because the notion and practice of 'development' were by then being used mainly to promote the interests of the digital tech sector, what I then termed '*Development for ICTs*' (D4ICT). It also had a much more explicit and rigorous theoretical framing, grounded in the Critical Theory of Jürgen Habermas (1974; 1978). Those who relish theoretical discourse would do well to read that book before embarking on this one, since many of the arguments of the present book are premised on those developed at much greater length in *Reclaiming ICT4D*.

Our thoughts and practices all vary depending on our cultural contexts and the diversity of experiences we have gained over our lives. We all look at the world slightly differently, but powerful truths can emerge when the ideas of people from very different backgrounds start to coalesce. I also believe strongly in the power of stories to convince and persuade. Hence, a central feature of this book is its incorporation of vignettes written by 31 authors, who have a rich diversity of experiences in using digital technologies with some of the world's poorest and most marginalised peoples. These stimulating and motivating contributions add richness and colour to the central arguments of the book, and I am immensely grateful to everyone who has contributed so enthusiastically to this adventure.

In a few places in the text that follows, the names of individuals and organisations have been anonymised to maintain privacy and the safety of authors. Every effort has, however, been made to ensure the authenticity of all claims, and references to other sources where they exist have been provided in the bibliography. I use the words 'we', 'us', and 'our' liberally in the text in two main ways: first, to refer to all those from many different backgrounds who are committed to helping the poorest and most marginalised use digital tech in their own interests; and, second, to refer to you as the reader, and me as the author.

Above all, I hope that the arguments laid out in this emancipatory manifesto will encourage readers to look again at their own engagement in using digital tech in 'development' in new, more self-critical ways, and that, as a result, we can all commit to working together more responsibly with the poorest and most marginalised to enable them too to reap the benefits of using digital technologies safely, wisely, and securely in their own interests. This is an urgent task in an increasingly volatile world.

Acknowledgements

Many people and experiences have shaped the ideas that have emerged in this book, and I am so grateful to all of the authors who have contributed such powerful vignettes. It has been a real pleasure to weave this tapestry together with them all. I am also especially grateful to the following people who took the time to read and comment on parts of it while it was being written (listed in alphabetical order of family name): Piroska Bisits Bullen, Yu Ping Chan, Lieselot Declercq, Pari Esfandiari, G. 'Hari' Harindranath, David Hollow, Janet Longmore, Ariane Racine, Jake Unwin, and Chris Yates. Their inputs have considerably strengthened its arguments and have hopefully helped to make them more accessible to and relevant for the reader. I have also benefitted from many conversations and responses from others to specific questions, and I would particularly like to thank Shabnam Aggarwal (on entrepreneurship and the me-syndrome), Ken Banks (on entrepreneurship and innovation), Angelika Bullinger-Hoffmann (on the Innovation Mob), João Carvalho (on e-Government), Sameer Chauhan (on the UN system and UNICC's work), Anders Frigaard (on entrepreneurship), Torbjörn Fredriksson (on UNCTAD), Ralph Hertwig (on many of the psychological aspects of the book), Poncelet Ileleji (on innovation hubs and digital sovereignty), Mark Mulobi (on the Digital Principles for Development), Matt Scullin (on the origins of 'personal computing'), Vladimir Strankovic (on WSIS), Mike Trucano (on innovation), Wayan Vota (on Failfaires) and Tom Wambeke (on innovation and training) for sharing their thoughts, knowledge, and experiences. Specific further thanks are also due to the following for permission to use some of their work for illustrative purposes: Ken Banks (Figure 5.2), the cartoonist Alok (Figure 5.5), Piroska Bisits Bullen (Figure 5.6), Marine Al Dahdah (Figure 5.7), and Emma Edwards (Figure 7.2). I am also very grateful to those who have taken their time to read much of the book in draft, and write such kind endorsements of it. Especial thanks are also due to Helena Hurd and Katerina Lade at Routledge who have provided considerable advice and support throughout the book's genesis and the preparation of the manuscript, as well as the very necessary encouragement that is so important at the difficult moments in crafting a book such as this. I am also extremely grateful to Susan Dunsmore for the wise and skilful copy-editing that she undertook, which has undoubtedly helped to make the book cleaner and more balanced, and to my production editor Jen Hicks, not least for spotting things that I had missed during the final stages of the book's preparation.

Above all, I am very deeply grateful to all those who have shaped my understandings of international development and digital tech since I first worked with Sudhir Wanmali in what was then South Bihar, now Jharkhand in 1976 and 1977. These include my colleagues in DFID in the early 2000s, and the out-of-school youth, people with disabilities, young women, teachers, and schoolchildren across Africa with whom we then worked. I am especially grateful to Alex Wong and other colleagues at the World Economic Forum in the late 2000s, and the many friends in UNESCO, especially Kōichirō Matsuura and Abdul Waheed Khan, but also Davide Storti, Dominique Morisse, Indrajit Banerjee, and Armelle Arrou-Bertrand, who over the last 25 years have been so generous with their time. Friends and colleagues at the ITU, especially Doreen Bogdan-Martin, Mario Maniewicz, Jose Maria Diaz Batenero, and Attila Horváth, have all taught me much, and I am also particularly grateful to Jasmina Byrne and Juan Pablo Giraldo Ospino at UNICEF, Torbjörn Fredriksson at UNCTAD, and Srinivas Reddy and Paul Comyn at the ILO for sharing so many valuable insights with me over many years. My staff at the Commonwealth Telecommunications Organisation in the mid-2010s taught me so much, and I am very grateful to them and all the Ministers and officials from across the Commonwealth who were our members at that time for their many insights and advice. Within the various British governments and Civil Service over the last 25 years I have learnt much from Hilary Benn, David Lammy, Sarah Taylor, Nigel Hickson, and Jamie Saunders, all of whom have supported me at critical times in my endeavours. I have learnt at least as much, though, from the very many economically poorer and less privileged people across the world whom I have sought to serve, from members of the LBT community in the favelas of Rio de Janeiro, to migrants from Nepal and refugees in South Africa; from out-of-school youth in Ethiopia to people with disabilities in Pakistan; from women at threat of sexual violence, especially in powerfully patriarchal societies, to African teachers eager to learn how to use digital tech to enhance their teaching; and from schoolchildren in rural China, to young entrepreneurs in the Caribbean. It is their voices above all else that have motivated me to craft this book, and I hope that it does justice to all that they have taught me.

Abbreviations and Acronyms

ADHD	Attention Deficit Hyperactivity Disorder
AGI	Artificial General Intelligence
AI	Artificial Intelligence
APC	Association for Progressive Communications
Apps	Applications
AR	Augmented Reality
ASI	Artificial Superintelligence
BCG	Boston Consulting Group
BoP	Bottom of the Pyramid
BPO-ITES	Business Process Outsourcing-Information Technology Enabled Services
BRICS	Brazil, Russia, India, China, South Africa
BRICS+	Brazil, Russia, India, China, South Africa, Iran, the United Arab Emirates, Saudi Arabia, Argentina, Egypt, and Ethiopia
CATIA	Catalysing Access to ICTs in Africa
COL	Commonwealth of Learning
CRS	Catholic Relief Services
CEB	Chief Executives Board for Co-ordination
CSC	Common Service Centre
CSER	Corporate Social and Environmental Responsibility
CSO	Civil Society Organisation
CSR	Corporate Social Responsibility
CTO	Commonwealth Telecommunications Organisation
CX	Customer Experience
D4ICT	Development for ICT
DESC	Digital Environment System Coalition
DFID	Department for International Development
DIAL	Digital Impact Alliance
EMI	Equated Monthly Instalment
FAO	Food and Agriculture Organization
FCDO	Foreign, Commonwealth and Development Office
FLOSS	Free/Libre and Open Source Software
FOSS	Free and Open Source Software

GBV	Gender-Based Violence
GBViE	Gender-Based Violence in Emergencies
GDC	Global Digital Compact
GDPR	General Data Protection Regulation
GSMA	GSM Association
HCI	Human-Computer Interaction
HLCP	High-Level Committee on Programmes
HLPDC	High-Level Panel on Digital Cooperation
IAEA	International Atomic Energy Agency
IBRD	International Bank for Reconstruction and Development
ICANN	Internet Corporation for Assigned Names and Numbers
ICT4D	Information and Communication Technology for Development
ICTD	Information and Communication Technology and Development
ICTs	Information and Communication Technologies
IDB	Inter-American Development Bank
IDRC	International Development Research Centre
IEA	International Energy Agency
IEEE	Institute of Electrical and Electronics Engineers
IGF	Internet Governance Forum
ILO	International Labour Organisation
ILOAT	Administrative Tribunal of the ILO
ISOC	Internet Society
ITCILO	International Training Centre of the ILO
ITU	International Telecommunication Union
LGBTIQ	Lesbian, Gay, Bisexual, Transgender, Intersex, and Queer or Questioning
LGBTQ	Lesbian, Gay, Bisexual, Transgender, and Queer or Questioning
LGBTQI	Lesbian, Gay, Bisexual, Transgender, Queer or Questioning, and Intersex
LMICs	Low- and Middle-Income Countries
M4D	Mobiles for Development
M4Dev	Mobiles for Development
M&E	Monitoring and Evaluation
MAG	Multilateral Advisory Group
MDGs	Millennium Development Goals
MERL	Monitoring, Evaluation, Research, and Learning
NGO	Non-Governmental Organisation
ODA	Official Development Assistance
ODET	Office for Digital and Emerging Technologies
ODL	Open and Distance Learning
OER	Open Educational Resources
OHCHR	Office of the United Nations High Commissioner for Human Rights
OTT	Over-The-Top
PIT	Public Interest Technology
SDC	Swiss Agency for Development and Cooperation
SDGs	Sustainable Development Goals

SG	Secretary-General
UAV	Unmanned Aerial Vehicle
UDHR	Universal Declaration of Human Rights
UN	United Nations
UNCED	United Nations Conference on Environment and Development
UNDESA	United Nations Department of Economic and Social Affairs
UNDP	United Nations Development Programme
UNECE	United Nations Economic Commission for Europe
UNESCO	United Nations Educational, Scientific and Cultural Organization
UNGA	United Nations General Assembly
UNHCR	United Nations High Commissioner for Refugees
UNICEF	United Nations Children's Fund
UNIN	United Nations Innovation Network
UNODET	UN Office for Digital and Emerging Technologies
UNOSET	UN Office of the Secretary-General's Envoy on Technology
UNSD	United Nations Statistics Division
USAF	Universal Service and Access Funds
USAID	United States Agency for International Development
USF	Universal Service Funds
WFP	World Food Programme
WHO	World Health Organization
WISE	World Innovation Summit for Education
WRC	World Radiocommunication Conference
WSIS	World Summit on the Information Society
WTDC	World Telecommunication Development Conference
WTO	World Trade Organization
WWW	World Wide Web

1 Notes from the underground

Tim Unwin

> You, for instance, want to cure [people] of their old habits and reform their will in accordance with science and good sense. But how do you know, not only that it is possible, but also that it is *desirable* to reform [people] in that way? And what leads you to the conclusion that [someone's] inclinations *need* reforming? In short, how do you know that such a reformation will be a benefit to [them]? And to go to the root of the matter, why are you so positively convinced that not to act against [their] real normal interests guaranteed by the conclusions of reason and arithmetic is certainly always advantageous for [them] and must always be a law for [humanity]?
>
> (Dostoyevsky, *Notes from the Underground,* 2021, 1: IX; first published, 1864)

The underground

Digital technologies have become increasingly prominent in the lives of most – but not all – people over the last quarter of a century. Whose interests does this really serve? Are these technologies going to make the world a 'better' place, especially for the world's poorest and most marginalised peoples? Or are they instead going further to enslave and exploit them, incorporating them into a system whereby they become increasingly unfree, with the 'Digital Barons' extracting ever more surplus value from the data of their day-to-day lived experiences? This book is about understanding why the world's poorest and most marginalised are being systematically disadvantaged through the use of digital technologies, and what our responsibilities are if this is to change. The epigraph from Dostoyevsky opens the door not only for this chapter but for the entire book. It builds on my own reflections (Notes) on working in digital tech and development, often in some of the darkest spaces (the underground, or 'crawl space'), between the cracks, and across boundaries. It challenges utopianism, but in the context of rhetoric about digital tech and development rather than mid-nineteenth-century Russia. More specifically, the quotation above introduces a critique of scientism and determinism which is central to the arguments that follow. *Notes from the Underground* (variously translated) was also published just after the abolition of serfdom in Russia in 1861, and provides interesting reflections on the potential for eliminating contemporary digital enslavement.

DOI: 10.4324/9781003598008-1

Think about how you use your own digital devices. Those making and selling each new 'phone you buy claim that it will bring you improved services, will be faster, will have better cameras and memory, and will have Artificial Intelligence (AI) as an integral feature. By definition and design, new devices are intended to be 'better' for you. But are they? Could you survive without them? What about all those who only have an old second-hand device, or do not even own a 'phone at all? By purchasing your new 'phone, are you not increasing your 'distance' from them; your world experiences are becoming more unequal. At the same time, you are also contributing to the profits of the global digital tech companies who make the devices, build the infrastructure, run the networks, and provide the apps that run on them. You are enabling those companies as well as governments to surveil you to an ever greater extent in their own interests. In this light, might not those without such devices actually have an advantage over you? Might not the right to be unconnected be more important than a putative right to be connected?

This book explores these issues by drawing on the diverse experiences of many people with extensive research practice who have been directly involved in delivering interventions using digital tech in the interests of the world's economically poorest and most marginalised people. Many of these experiences are captured in the vignettes and slightly longer spreads between chapters, which illustrate and sometimes challenge the arguments contained in the main body of the text. They provide real-world examples of what has worked and what has not, and many also adopt a normative stance, suggesting what needs to be changed if digital technologies are really to be used to make the world a 'better' place for all. As the chapters evolve, they present an increasingly dystopic view of the future, but ultimately the book has a positive message. Digital technologies need not be constructed and used as they currently are, but achieving this will require a truly revolutionary transformation of the systems and structures that dominate today's increasingly digital world.

Digital Inclusion in an Unequal World has a practical intent in sharing evidence of good practices in order to influence future behaviours, while also shedding light on the underlying interests that have led to the existing structure of digital tech, development and inequalities. There have undoubtedly been some successes in using digital tech to serve the interests of the world's poorest and most vulnerable over the last quarter of a century, but why have the lessons from existing good examples been insufficiently understood or drawn upon by those implementing new, 'innovative' initiatives? Why do people take pride in claiming that they are 'building the plane as they fly it', while also asserting that they are creating 'new' digital tech for development initiatives, without ever having studied in detail previous attempts to do the same thing? Reinventing the wheel is indeed sometimes necessary, but far too often today it is done through ignorance and the avarice of individuals and organisations, rather than through wisdom and rigorous understanding.

This introductory chapter sets the scene by reflecting on four specific instances from my own experiences in the underground, i.e., the darker side of digital tech and development. These 'Notes' shed light on the most important reasons why digital tech has been used to increase inequalities, despite expansive global rhetoric around their positive benefits for all. They provide the basis for the arguments that are developed further in the subsequent chapter.

Note one: the global geopolitical framework

Digital technologies are now widely seen as being instrumental in delivering the UN-led Sustainable Development Goals (SDGs). The opening paragraph of the Global Digital Compact (GDC), agreed in September 2024, thus begins by claiming that 'Digital technologies are dramatically transforming our world. They offer immense potential benefits for the well-being and advancement of people and societies and for our planet. They hold out the promise of accelerating the achievement of the Sustainable Development Goals' (UNOSET, 2024, p. 1; see also Figure 1.1). Yet, as the figures below clearly indicate, there remain profound differences between the use of digital tech in different parts of the world, between men and women, and between urban and rural areas. Not everyone benefits equally from the use of digital tech.

In 2024:

- 32% of the world's population were not using the Internet; **in Africa, only 38% of the people were using the Internet**
- Globally, 65% of women and 70% of men used the Internet; **in low-income countries, 21% of women and 32% of men used the Internet**.
- Globally, 83% of urban people used the Internet; **only 48% of rural people did so.**
- In high-income countries, there were 138 mobile 'phone subscriptions per 100 people; **in low-income countries there were only 71**.
- Data-only mobile-broadband (2GB) basket prices cost 0.4% of GNP per capita in high-income countries; **in low-income countries it cost 7.4%.**

(ITU, 2024)

Figure 1.1 UN Secretary-General António Guterres addressing the opening of the Summit of the Future, 2024

Source: UN Photo/Loey Felipe (2024).

Countless new initiatives have been, and continue to be, developed at the global and regional levels by organisations and individuals within the UN system that claim to be addressing these inequalities, but most are doomed to failure for two reasons: first, the SDGs and Human Rights agenda of the UN will never deliver successfully on the needs of the world's poorest and most marginalised peoples (see Chapter 3); and, second, private sector corporations have fundamentally subverted the UN agenda to serve their own interests, especially in the digital tech sector (Sneyd, 2024).

If anyone is in doubt about these claims, consider just the creation and practice of Giga, an initiative by the United Nations Children's Fund (UNICEF) and the International Telecommunication Union (ITU) that is designed to connect every school to the Internet and every young person to information, opportunity and choice. On the face of it, surely these actions have to be positive? Education is increasingly being delivered through the Internet, and there are many who argue that access to it is now a basic human right (Howell and West, 2016). Surely it can only be a good ambition to connect all schools everywhere to the Internet? However, such initiatives also increase the market for digital tech companies, and especially those in the so-called EdTech sector. Having mapped all the world's schools, Giga's core activities are primarily designed to increase the financing available for such a roll-out. As Giga claims, it is 'mobilizing $5 billion to accelerate the delivery of critical infrastructure and close the digital divide' (Giga, n.d.). In effect, this is primarily serving to reduce the costs of entry into difficult markets for private sector companies. Moreover, by focusing on education alone, is it not also in danger of duplicating efforts by others to use the Internet for benefits such as health and the provision of online government services? What is really needed is surely an integrated state-led approach to Internet for all, everywhere.

So, why are UN agencies so eager to develop such initiatives that are destined not actually to serve the interests of the poorest and most marginalised? After all, the most marginalised are usually those who are not even in school, so connecting schools to the Internet will only further marginalise the already marginalised. To answer such questions, it is necessary to enter into the murky waters of individual career paths, competition between UN agencies and their senior management teams, and the methods used by the private sector to influence global policy through the UN system. These issues are all pursued further through the remainder of this book, and especially in Chapter 3.

Note two: the 'me' syndrome

In 2019, I sat in a large UN forum listening to a keynote in which the presenter shared their experiences of creating a digital tech initiative that they claimed would transform education and make the world a better place. They were in their thirties, and emphasised their own success in previously working in government, as a barrister, and in a start-up company. They had won awards and gained status, and thus seemed to feel entitled to use their experience to have an impact on poverty reduction. They spoke eloquently about the potential of AI to transform teaching and learning. Many were transfixed by the smooth presentation. However, some who had been working on digital tech and education for many years thought otherwise. They saw this primarily as an entrepreneurial pitch to boost the reach of the company that the presenter had founded six years previously. The speaker seemed completely unaware of the many previous interventions that had tried and failed to deliver the educational outcomes that they were also hoping to achieve. Many applauded that person. It was a triumph of rhetoric over reality.

I reflected long and hard on this. The presenter undoubtedly had skills, enthusiasm and a strong work ethic, but seemed to have made little effort to examine what had gone before, and learn from previous mistakes. They saw technology as an unquestionable force for good – and for profit. Why was this? If only their commitment could have been matched by wisdom. Then something might really be achieved for the world's least educated, poorest and most marginalised people.

The above story could be told many times over in very different contexts (see vignettes by Nwakanma, p. 16, Hollow, p. 126 and Sterling, p. 66). It is the entrepreneur's pitch (Banks, 2013a; 2016). A competent person, caring passionately about what they are doing, and believing that they really can make the world a better place, sets about creating a new initiative with the stated intention of using digital technologies to deliver on a particular aspect of 'development'. Speaking at a UN event, and linking their work to the SDGs is an ideal vehicle through which they can claim to be doing 'good'. In many instances, these visions are well-intentioned, although that is by no means always the case. Frequently, they are carefully crafted, planned, and implemented with steely precision in the interests primarily of making profits and further building their reputations and bank balances. Sometimes it is only the rhetoric that matters.

Although the underground Note above focuses on an individual, this same logic applies to many civil society organisations, start-up companies, global corporations and even governments and international agencies engaged in the use of digital tech for development. The rhetoric is far too often more important than the reality. However, these entities are not necessarily immutable monolithic structures that cannot be changed. They are made up of myriads of human beings, each with their own ambitions and desires. Some undoubtedly along the way do 'good', in a sense that most people would recognise it, but frequently they insufficiently address the real needs and interests of the world's poorest and most marginalised people. Why is this? Why do initiatives ostensibly designed to reduce poverty and improve the lives of the world's most marginalised people so rarely achieve the outcomes to which their rhetorics aspire?

In part, their low success rate is because of deep structural reasons. The dominant neoliberal global system until recently led by the USA, focusing on the capitalist requirements of expanding markets and reducing labour costs, is undoubtedly part of the explanation. This system requires poor people from whose labour surplus value can be extracted and converted into profits by the owners of capital. It also requires expanding markets so that this profit can be realised. It is no coincidence that until very recently most of the largest digital tech companies in the world have been located in the USA. Even in 2024, the six largest tech companies by market capitalisation (Apple, NVIDIA, Microsoft, Alphabet, Amazon, and Meta) were all still located there (Companiesmarketcap, 2024).

However, such structural constraints are not the only reason. There are very strong personal factors that also need to be understood and redirected if there is to be any fundamental shift towards digital tech being used appropriately by poor and marginalised people to improve their lives. Individual greed, ambition, and wealth are powerful driving forces that have shaped today's world of digital tech. It is essential to understand the interests underlying the ways through which these technologies are designed, produced, marketed, and used, if there is to be any change in their impact on the lives of the economically poorest and most marginalised. There needs to be a moral reawakening in the tech sector (see also Seth, 2022).

It must be recognised, in particular, that there is no such thing as value-free science. Those who have conceived and designed new digital technologies have done so for particular reasons, and reflect specific interests. Scientists are as responsible for the unintended consequences of their work as they are for the intended ones. There is also nothing 'natural' about technological 'advances'. The World Wide Web and the Internet would not have been as they are today if they had not been developed in the way that scientists, such as Berners Lee (2025) and Cerf, conceived them; Apple would not have been Apple without Jobs, Wozniak, and Ive; Microsoft would not have been Microsoft without Gates (Schwab, 2023); Amazon would not have been created without Bezos (Brandt, 2011); Alphabet, the parent company of Google, would never have existed in the way that it is now without Page and Brin; and SpaceX would not exist as it is without Musk (Arthur, 2012). None of these people placed much emphasis on reducing global inequalities and directly serving the needs of the economically poor and marginalised when they began designing their technologies and building their companies. Indeed, the prime drivers for many of them were the enjoyment that they got out of 'making tech', and the potential this could offer for them to become rich. They were not driven by a passion for reducing inequalities and serving the needs of the world's marginalised, although the market opportunity that 'international development' now offers definitely contributes to their companies' ongoing bottom lines (Unwin, 2017a). Individuals who work for companies such as these have choices; is it possible to serve their corporate interests while also creating more equitable societies? Is it possible to do good whilst making vast fortunes?

It is also important to note that all the people cited in the above paragraph are white men. What if these 'ground-zero' digital technologies and companies had been created by black or brown women? Why was that apparently never even a possibility? Many women were certainly involved in the early days of digital tech, and others continue to hold important roles in the industry, but might it have made a difference if such global corporations had been created and led by them?

These complex issues rise to the forefront in many parts of this book, but are addressed in specific detail in Chapter 4 which explores how the 'me' syndrome operates at different levels, from individuals, through companies and civil society organisations, to the wider international arena. The world of digital tech does not have to be as it is; it has been shaped by people who have very specific interests, and who mostly know very little about poverty and the lives of the most marginalised. Throughout history, those in power have created and used technologies above all else to maintain their positions of power. However, because all individuals inherently do have some free will, I retain the hope that enlightened and emancipated people might be able to create a different kind of world where digital technologies are indeed used to serve the interests of the world's poorest and most marginalised people.

Note three: corporate explorations in 'darkest' Africa

In the late 2000s, I was sitting in a small meeting room in a conference hotel discussing the roll-out of the World Economic Forum's Partnership for Education initiative with UNESCO (Unwin and Wong, 2012). There were leaders of industry and government officials present, and as so often things didn't always run according to the time schedule. This provided much opportunity for interesting side discussions. Sitting next to me was a senior manager of a global tech company that had been involved in the NEPAD e-Schools Program in Africa (Farrell, Isaacs and Trucano, 2007). I took the opportunity to enquire about his involvement, noting that, to my mind, it had been largely a failure as far as its educational outcomes were concerned. To my initial surprise, he fundamentally disagreed and argued that it had been a great success. I naïvely looked quizzical, suggesting that surely it was the educational outcomes that mattered most. He said, not at all, what mattered most for his company was the access that it gave them to senior government figures in potential markets, and to like-minded organisations with whom they might develop synergistic partnerships. I remain very grateful to him for opening the door to understanding the interests behind the explorations of digital tech companies in 'darkest' unknown Africa.

Private sector companies have a fundamentally important role to play in designing, producing, and selling goods and services for the benefit of humanity. It is largely the technical innovations of such companies and individual entrepreneurs that have made the world a very much better place to be living in today than heretofore (Rosling *et al.*, 2018). This is especially true of digital tech companies and their predecessors in the telecommunications sector, which had historically maintained close relationships with governments, not least through their joint participation in international organisations such as the International Telecommunication Union (ITU), founded in 1865, and the Commonwealth Telecommunications Organisation (CTO), founded as the Pacific Cable Board in 1901. International agreements have been essential for enabling global telecommunications and digital infrastructure to be constructed and maintained. However, with the privatisation of the telecommunications sector from the 1980s onwards (Megginson, 2005), an entirely new dynamic was created. No longer were telecommunications seen primarily as being a service for all citizens provided by their governments, but rather as an opportunity for private capital to invest in businesses across the world and from which substantial profits could be generated. The profit motive, rather than primarily a service model, has enabled the creation of a vast swathe of new digital technologies, but it has come at a cost. That cost has been the potential for inequalities between rich and poor to be increased, both between countries and within them.

Global inequalities declined in the latter part of the twentieth century, but are now at least as high as they were at the peak of European imperialism. As the World Inequality Lab report (Chance *et al.*, 2022, p. 11) notes, 'Income and wealth inequalities have been on the rise nearly everywhere since the 1980s, following a series of deregulation and liberalization programs which took different forms in different countries', but as they go on to say, 'The rise has not been uniform'. Again, this offers a glimmer of hope, because it reaffirms that inequality is not inevitable but rather reflects differences in economic, social, political, and cultural choices. It also highlights the crucial importance of regulators in negotiating the balance of power between companies and citizens, a key topic to be returned to in Chapters 2 and 7.

The attitudes and behaviours of individuals working in private sector business, as well as the corporate strategies that they inform, are critically important in shaping the future of inequality and international development (Nelson *et al.*, 2005). Unfortunately, most senior managers in digital tech companies have extremely little knowledge of real crippling poverty, or of the work of governments and civil society organisations in the economically poorer parts of the world to try to alleviate this. For many of them, countries in Africa were largely unknown in the era of the UN Millennium Development Goals (MDGs) from 2000–2015; in that sense, they really were 'dark' (see also Unwin, 2008). Poor countries and people are primarily now seen as a market to be 'conquered' or 'penetrated' (on the use of such language, see Unwin, 2022a). It is not accidental that poorer *countries* have become known as 'emerging markets' or 'emerging economies' by those in the private sector and the wider bodies that they inform (see, for example, World Bank, 1981; World Economic Forum, 2024). This challenge is compounded when they define 'development' itself primarily in absolute terms as 'economic growth', a view that provided the underlying ideology for the MDGs and their successor SDGs (Unwin, 2007). In such contexts, it is scarcely surprising that digital tech companies and their staff focus primarily on seeking to increase their profits through expanding their markets and reducing labour costs, just as the owners of capital have always sought to do. It would be illogical for them to do otherwise (although see vignette by Hughes, p. 115).

This profit-seeking motive can, though, be co-opted into broader international development activities through effective multi-sector partnerships that combine partners from government, civil society, and the private sector (Unwin, 2015a). Such partnerships have become a central feature of many global initiatives involving digital tech in 'development', especially within the UN system, and are discussed in more detail in Chapters 2 and 3. At this stage, though, it is essential to emphasise that such partnerships only stand a very limited chance of success in reducing poverty and inequalities unless they are truly transparent, are underlain by a rigorous ethical framework, and are not dominated by the avaricious power and ideology of the major digital tech companies. Otherwise, as the note that introduces this section emphasises, this rhetoric of partnership is merely an illusion that will further subvert the good intentions of many of those who work within governments and the UN system.

Note four: just because you can do something does not mean you should do it

There's something very physical about digital tech and science that I love. In part, it is the binary certainty of 1 and 0, right and wrong. In part, it is the smell of machines, the oil, the cold metal of the 1970s, and the brightness and heat of energy. In part, it is the purity of mathematical proofs and beautifully simple code. But I have always railed against 'scientism', the belief by many scientists that science and the scientific method are the only way to render valid truth about reality and the world (Unwin, 1992; 2017a). This was vividly brought home by the Open Letter to the United Nations in July 2024 by a group of 37 scientists outlining their fears about the increasing regulatory controls being advocated in the draft text of the UN's Global Digital Compact (UNOSET, 2024). They sought 'to ensure that proposals for digital governance remain consistent with the enormously successful multistakeholder Internet governance practice that has brought us the Internet of today' (Applequist et al., 2024). Most of the scientists had worked with, or still contribute to private sector companies, and nearly all were men. While they are willing to take the credit for the success of the Web and the Internet, they do not fully accept that they are also responsible for the harms for which these technologies have been used. 'There is nothing inherently "good" about science; it serves particular sets of interests. Scientists are as responsible for the harms, unintended or deliberate, caused by their inventions as they are for any "good" for which they are used.'. . .'Did the engineers and others involved take the time to consider the full implications of what they were doing?' (Unwin, 2024a). If the Web and the Internet had been designed differently, might we have avoided some of the many harms for which they are now used?

Most people are in thrall to science and scientists. Scientists and engineers get things done; they produce the innovations that make the world 'better'; they provide firm evidence upon which decisions can be made. Scientists can also become very rich and successful. Across the world, there are hundreds of thousands of young scientists who wish to become app developers or coders, and emulate their heroes such as Musk, Gates, or Brin. Few if any, though, will achieve this. Moreover, science cannot produce answers to our most difficult questions. As we undertake more novel science, the old truths of science are disproven; science cannot ever claim therefore to provide absolute truth. Science is always an approximation. Most importantly of all, it cannot say whether what scientists do is good or bad. It lacks a moral compass. Scientists may claim to be value-free, and yet they are imbued with their own ideologies, epistemologies, and socio-economic interests. Even in universities, many scientists are primarily driven by the need to get large research grants and publish in the most esteemed journals (Harris, 2016). In the private sector, the need to build things, both real and virtual, to generate profits overrides almost everything else.

Hence, it is crucial that effective systems of accountability and responsibility are in place to ensure that the digital technologies that are brought to market are indeed designed to do good and to mitigate harm. It is insufficient simply to try to put regulatory systems in place after the event. Scientists in both the private sector and universities must be trained and subsequently required to consider the difficult moral questions about what they are doing (Seth, 2022). Rigorous processes need to be implemented across the world whereby scientists are always made to consider the potential unintended consequences of their work. Alfred Nobel came deeply to regret creating dynamite when it began to be used in warfare, but by then it was too late. Just because it is possible to do amazing things with the latest digital technologies, such as those that fall within the umbrella term Artificial Intelligence (AI), does not mean that it is wise or good to do so. Ask yourself why we need AI and cyborgs, and then reflect on your answers. Whose interests do they serve? Do they really serve the world's poorest and most marginalised?

The subversion of the UN system by private sector companies has already been alluded to at the start of this chapter, but this has been reinforced by its coalition of interests with scientists. Science adds truth claims and apparent validity to the discourse. This coalition is unsurprising, because much of the best and most innovative science in the world is now being developed in companies rather than universities, and this is especially striking in the digital tech sector. Most of the initiatives and boards created to consider issues around digital tech and development under the auspices of the UN Secretary-General have been heavily dominated by scientists and leaders of industry. Almost all of the Board members of the High-Level Panel on Digital Cooperation, created in 2020 (UN, 2020), including government officials, had experience working in digital companies, and were scientists, or had scientific backgrounds. Very few had any real understanding of the complexities of international development, or had backgrounds in the social sciences or humanities. Even fewer had by any stretch of the imagination ever been financially very poor. Likewise, although a very few members of the Secretary-General's AI Advisory Board, established in 2023 (UN, 2023a; 2024a), did have legal or ethical expertise, most, once again, were from scientific backgrounds.

In practice: reading this book

The above Notes provide a contextual introduction to the key themes of the book. The next chapters tease out the five most important inter-related issues that require resolution if the poorest and most marginalised people are not to be further disadvantaged by the dominant global digital tech agenda. The vignettes and slightly longer accounts in chapters are included to illustrate the arguments from a wide diversity of perspectives, and are mostly written by people who have experience both as thinkers and also as practitioners in using digital technologies to serve those less privileged than themselves. The first of these shares my own experiences in one of the earliest explicit partnerships that used digital tech for development in Africa in the early 2000s. It highlights many lessons that remain as relevant today as they were a quarter of a century ago.

Lessons from Imfundo: Partnership for IT in education, 2001–2004

Figure 1.2 Children with whom Imfundo worked in Ethiopia, 2002. This photograph was also used on the front cover of DFID's (2004) *The Rough Guide to a Better World and How You Can Make a Difference*

Source: Tim Unwin.

Imfundo (the isiXhosa and Zulu word for 'education') was an innovative UK Prime Ministerial initiative, based in the Department for International Development (DFID), that was created in 2000 to engage the private sector in delivering beneficial transformations in education in Africa through the use of Information and Communication Technologies (ICTs). Its inception during 2000 owed much to the work of Owen Barder, who had been Private Secretary to the Prime Minister for Economic Affairs and then moved to DFID as Chief Economist for Africa and Head of the Africa Policy Department. The inception established a basic structure for Imfundo: the secondment of one person each from Marconi, Virgin, and Cisco; the creation of its three core elements, a ResourceBank, a KnowledgeBank, and activities in Africa; pilot activities in Rwanda and The Gambia; and publication of 16 initial KnowledgeBank papers. In 2001, I was invited to lead Imfundo, based within DFID, and we subsequently created a diverse team in which three secondees and some 22 DFID staff members contributed at various times. Much of our work on the ground was delivered by Caroline Pontefract, our education advisor, and Bas Kotterink, our ICT adviser, but every member of our diverse team contributed to Imfundo's success.

Imfundo was one of the first explicitly multi-sector initiatives in the newly established field of ICT4D that sought to 'do development' through radically different means. Its vision was 'To create partnerships to contribute to the delivery of universal primary education and gender equality in Africa through the use of ICT', and it was innovative in two main ways: it sought to create a partnership model at a time when DFID's support for Africa was increasingly focused on budget support mechanisms; and it aimed to use ICTs *for* Education, when many did not even know what ICTs were, let alone believed that they could contribute to African education. I will never forget what a senior education advisor in DFID told me at that time: 'Children in African schools need pencils and paper rather than computers.'

Imfundo had three key principles: (1) it worked in support of DFID country programmes, despite being a central headquarter initiative; (2) it advocated for the appropriate and sustainable use of ICT *for* education, exploring the potential of multimedia in an African context, and emphasising the key role of teacher training; and (3) it placed emphasis on process not product, helping African governments to own the strategies, building African capacity, not least through Free and Open Source Software (FOSS), and avoiding the failures of inappropriate 'computers in schools' initiatives. Ultimately, we created a partnership network of around 40 entities from the private sector, civil society, and international organisations, and worked in eight African countries. However, at the end of four years, DFID chose to close Imfundo in favour of other initiatives, such as CATIA. This was in part a result of the partial and problematic review of ICT for poverty reduction undertaken for DFID by Marker *et al.* (2002). It also, though, reflected the considerable antipathy towards our objectives that existed at the time amongst other influential people within DFID. As Lowcock and Dissanayake (2024, p. 136) comment:

> [Imfundo] was regarded with some scepticism by DFID's education specialists, and by managers in DFID's country offices who thought that the main chance was in operating at large scale with governments to expand traditional systems, and who worried about being diverted from that.

We made available all our information and resources online, in hard copy and in regular newsletters, and some of the outputs have also been published (for example, Selinger, 2001; Unwin, 2004; 2005a; 2005b). However, despite our best efforts to share good practices and lessons learnt, countless subsequent ICT4D initiatives over the last quarter of a century seem to have started again from scratch without drawing on the groundbreaking work in which we, and others, were involved more than 20 years ago. One of the underlying purposes of this book is to try to understand why this is so.

With hindsight, six key conclusions that are important for the wider arguments of this book can be drawn from the work of Imfundo:

- *Partnerships*: these can be very effective, but need careful crafting and management (Unwin, 2015a). I am certain that our private sector secondees were well-intentioned, but it is clear that many companies today see their contributions to 'development' primarily as a means to increase their visibility and sales.

- *Power*: mind your back! The way Imfundo was structured required us to work very much through the DFID country offices, and I am conscious that initially I did not sufficiently do enough to ensure we had support from the highest levels in the 'office' (the headquarters). Ultimately, once we had started delivering effectively, we did win round the Secretary of State, Clare Short, but I should have done this earlier.
- *Teacher training*: we were passionate advocates of the importance of providing teachers with sufficient training before digital tech was rolled out into schools. I cannot understand why this truism is so often ignored to this day.
- *Appropriate technologies*: it is essential to use appropriate, affordable, and sustainable technologies if the poorest and most marginalised are to benefit from them. This is why Imfundo placed emphasis on FOSS and thin-client systems in schools.
- *Effective monitoring and evaluation*: this is essential for any development initiative, but especially in the context of digital technologies. Without it, there is no way of knowing whether your impact is positive or negative, and unintended consequences must also be considered.
- *Ethics and altruism*: all digital partnerships must be grounded in a sound ethical basis. At the time Imfundo was founded, much was being said about the government's new ethical foreign policy (Wheeler and Dunne, 2004). When I naïvely asked for a copy of this ethical policy, I discovered that no such thing existed in DFID, and so we wrote an ethical framework for our partnerships that enabled us to decide which organisations we were willing to work with.

Despite its short life, the lessons from Imfundo have lived on in the work of many of those who were involved in it. Caroline Pontefract went on to be Chief of Teacher Education at UNESCO, Director of the Social Transformation Division at the Commonwealth Secretariat, and then Director of Education at UNRWA; Bas Kotterink became Senior Scientist at Toegepast Natuurwetenschappelijk Onderzoek. The seeds sown in Imfundo also grew into the creation of the ICT4D Collective and the UNESCO Chair in ICT4D at Royal Holloway University of London (2007–2024) the work from which has fed directly into the arguments of this book.

The following quotations from Imfundo's output to purpose review reflect the views of those we worked with in Africa and stand as a testimony from more than 20 years ago as to the success of what we sought to do:

- Imfundo's emphasis on working with the poorest of the poor communities and collaborating with black SMMEs is exemplary.
- The best networkers in the donor community.
- Imfundo is the only member of the international community that has shown any interest in us (a black start-up company).
- Strategic partnership approach and transparent, participative, networking processes are excellent, as is direct involvement of Imfundo with relevant expertise. Most other donors don't work like this.

The remaining chapters build on the experiences described above. They shed light on why, despite there being much good evidence over a long period on how to deliver digital interventions that can indeed benefit the poorest and most marginalised, they usually do not have such outcomes (Weigel and Waldburger, 2004; Milward-Oliver, 2005; Elder *et al.*, 2013).

Chapter 2 examines the extent to which a development agenda focusing on economic growth is compatible with notions of inequality and equity. It directly addresses the inequalities driven by the interests of private sector digital tech companies seeking to expand their markets and drive down labour costs. It concludes that regulation in the interest of the many rather than the few should be an urgent priority for governments that seek to serve all their citizens.

Chapter 3 examines how these private sector companies have increasingly sought to shape and influence global decision making about digital tech, especially through their partnerships with UN agencies, and more recently directly through the UN Secretary-General's office as described above in Notes One and Four (see also warnings by Martens, 2007). It argues that the SDGs have already failed, and that the UN should now be looking to the future beyond 2030 rather than trying to shore them up through the use of digital tech, which is only likely to make matters worse. Part of the reason for this is the problematic multistakeholder models that have been adopted within the global digital development community, and the chapter argues that using the term multi-sector instead can begin to change the imaginaries of those engaged in delivering digital tech for development interventions. The chapter also summarises the reasons why the Global Digital Compact (GDC) was a largely pointless exercise as far as the world's poorest and most marginalised are concerned.

Drilling down to the diverse actions and aspirations of the individuals and organisations involved in these discussions and interventions, Chapter 4 addresses the complex motivations behind engagement in digital tech for development, building on the observations in Note Two, but also those in Notes Three and Four, and the Imfundo case study. It explores practical issues such as the short-termism of project delivery, insufficient monitoring and evaluation, and the balance between individual and communal interests, arguing that the poorest will only benefit when the balance shifts from 'me' to 'we', a theme that recurs in many of the vignettes (see those by Nwakanma, p. 16 and Juliette Unwin, p. 85).

This in turn leads to a discussion in Chapter 5 around what I have termed the 'innovation fetish' that has come to dominate much discourse in digital tech and international development over the last decade. It is remarkable how each new wave of innovation, from mobile devices to Blockchain and AI, makes very similar claims about contributing positively to 'development', assuring us that they are the solution for delivering the SDGs. Yet those promoting them generally do not learn sufficiently from experiences of the past. In particular, the chapter challenges the notion that 'agile' approaches can provide a beneficial new approach to project implementation in the field of international development; for many of the world's poor, failure means even greater immiseration. We need to shift from innovation for growth to innovation that truly focuses on the needs of the poorest and most marginalised (see vignettes by Banks, p. 100, Harindranath, p. 96, and Hughes, p. 115).

The penultimate chapter, Chapter 6, is the most dystopic, and argues that we have already entered an era of digital enslavement. If slavery is defined as a loss of freedom, then the Digital Barons, mainly in the USA but also increasingly in China, have already enslaved most of those who are digitally connected, from whom they extract very considerable amounts of surplus value. If you cannot live without using your digital device(s) for a day, let alone a week, you have already become their slave. Unless changes are made very urgently to the ways in which the use of digital technologies for development are understood and implemented, all those of us working in the field are in danger of being complicit in this conspiracy. The final chapter, Chapter 7, pulls these ideas together into an emancipatory programme for the future. Despite the challenges, it finds hope in human free will and the opportunity to rediscover that living vibrantly in the real world is immeasurably better than being enslaved in a virtual world over which we have increasingly little control.

Working WITH, not FOR

Nnenna Nwakanma

(God-ward, Africa, online, policy, digital, cooperation, open, strategy, data, ambassador)

It is 2025. I am in the capital city of one of West Africa's 'forward-looking' countries in terms of digital development, working with the government and other major stakeholders on Internet connectivity. I listen to a lot of talk and I read countless reports about what is being done FOR the population. I make a request to go down to the school that the media has touted as the 'model'. I needed to see things for myself, where the rubber hits the road. I also refused any 'official' or 'media-related' accompaniment.

First shock: the school does not have a wall around it. There is no security, not for the teachers, not for the students, and most definitely not for the computers. However, the 'digital hub', which stood out because it was a new building, had iron doors and seemed to enjoy a level of security that other buildings in the school did not.

Second shock: The digital hub was under lock and key. I had expected that on a good day, students would be there doing things under the guidance of a teacher. When the doors finally opened, I got the biggest shock: all the donated equipment was unopened and in the same cartons and packaging that they had arrived in six months earlier.

I could not leave without understanding what went (was going) wrong. So, I sat down to listen. This is what I heard: a teacher is supposed to teach 'ICT – Information and Communication Technology' to the students. I asked the ICT teacher why classes had not started. The response: 'Madam, I was hired as a Mathematics teacher. I have my scheduled work-load all mapped out. If I am to start teaching ICT, I need to be trained for that and I need my bosses to reorganize my workload.'

I turned to the school administrator. 'Boss, what is going on?' The response: 'The teacher here is from the Ministry of Education. The Digital Hub was an initiative of the Ministry of ICT. As a school administrator, my employer is the Ministry of Education and I follow my employer's scheme. The Ministry of ICT did not say anything about employing and training teachers for the Digital Hub.'

Many semesters after the media launch, the Digital Hub in that school is still under lock and key:

- The teacher is waiting to be trained.
- The school administrator is waiting for the Ministries of Education and ICT to speak to each other.
- The school does not have the budget required to connect the Digital Hub to the Internet.
- The school is yet to have a fence, and connecting to the electricity grid is an issue yet to be addressed.

When are we going to stop working FOR stakeholders and start working WITH them? Why sit in nice air-conditioned places to make decisions FOR places that have neither basic security nor electricity? How many models exist only for media purposes? We've seen it all before: initiatives that are handed down with no intention of making any impact. Digital initiatives FOR that did not start WITH the supposed beneficiaries.

Stillbirths!

2 Shifting the balance

From growth to equity

Tim Unwin

> No man is an island, entire of itself; every man is a piece of the continent, a part of the main. If a clod be washed away by the sea, Europe is the less, as well as if a promontory were, as well as if a manor of thy friend's or of thine own were: any man's death diminishes me, because I am involved in mankind, and therefore never send to know for whom the bell tolls; it tolls for thee.
>
> (John Donne, 1624)

Donne's famous lines emphasise the fundamental relational argument of this chapter that we all have social connections with others, wherever we are in the world; none of us is separate from, or better than, the rest of humanity. This expresses itself in the need for all those engaged in digital tech for development to shift the balance of our interests towards social equity rather than economic growth. The poorest and most marginalised will never benefit sufficiently from the use of digital tech unless this happens. Donne's focus on death also highlights the reality that digital tech is increasingly being used explicitly to kill people, most visibly in the use of digitally controlled weaponry, and serves as a reminder that those developing these technologies are indeed diminished by such use. His emphasis on the relational is pertinent more widely to this book's discussion of partnerships in digital tech for development, and is central to Chapter 7 which returns in a very practical way to the responsibilities that we all have to others in using digital tech for development.

Private sector companies do not survive if they do not generate profits for their owners and shareholders. They are the engines of growth that have enabled immense improvements in people's lives across the world, but their core role is not to deliver goods and services to the world's economically poorest and most marginalised people who cannot pay for them. No-one should expect companies alone to bear the brunt of delivering services to them. Historically, this role has mainly been provided by governments redistributing wealth through taxation or by charitable giving primarily to civil society organisations. Hence, it is crucial for the private sector (including companies, foundations, and universities), governments and civil society organisations to work together in effective, responsibly-based partnerships if the most marginalised are to benefit.

DOI: 10.4324/9781003598008-2

The neoliberal agenda of the late twentieth century, with its focus on privatisation, deregulation, free trade, globalisation, and reductions in government spending, has nevertheless systematically undermined such partnerships and has instead primarily served the interests of global corporations and private sector companies, at the expense of the world's poor. This dominant neoliberal agenda has become embedded in the coalition of interests around economic growth that provided the basis for the Millennium Development Goals (MDGs) up to 2015, and subsequently also the Sustainable Development Goals (SDGs). This is despite Rodrik's (2017) warning that 'neoliberalism must be rejected on its own terms for the simple reason that it is bad economics'. Economic growth has lain at the heart of both the MDGs and the SDGs, and was clearly represented in Target 1A of Goal 1 of the MDGs: to 'Halve between 1990 and 2015, the proportion of people living on less than $1 a day' (UN, 2010). The crucial thing to note about this is that it was based on an *absolute definition* of poverty, and was in theory at least therefore quite easy to deliver. It did not explicitly consider *relative poverty* (O'Boyle, 1999; Unwin, 2007; 2017a), which is much more difficult to reduce. The reality is that economic growth raises the potential for relative poverty to increase; it widens the potential gap between the richest and the poorest. This is precisely what has happened in practice at all scales since the start of this century, and the digital tech sector, by focusing almost exclusively on the notion of 'development as economic growth', has been one of the prime agents of this increase in relative poverty (Chance *et al.*, 2022). It is no coincidence, for example, that SDG10 (on reducing inequalities) is almost always ignored in global discussions of digital tech. As but one indication of this, the International Telecommunication Union's (ITU) summary of how digital tech can contribute to every SDG has less written on SDG10 than about any of the other SDGs (ITU, 2021). The rhetoric that digital tech can solve the challenges of 'development' is a mirage for the poorest and most marginalised. The present system has not worked for them, and we must therefore seek alternatives if we wish to create a fairer and more equal world (see also Chapter 3).

This chapter explores the reasons why and how the digital tech sector has driven increased inequalities. It argues that there needs to be a fundamental repositioning of international practice at the interface between digital tech and development so that it focuses primarily on *equity* rather than *growth*, if we wish to be serious about serving the interests of the poorest and weakest in our societies. It begins by reflecting briefly on the diverse array of interpretations that exist about the notion of 'development', and defining the way in which key concepts are used in this book. This is crucial, because our language reflects our thoughts, and is one of the main ways through which we reach agreement and consensus. The chapter subsequently provides illustrations of how digital tech has been designed and used to increase inequalities at all scales. This discussion then leads into an examination of why economic growth is such a powerful rhetoric, and how the digital tech sector has exploited this in its own interests, especially through the development of new types of 'production' based on individual data. Finally, the chapter argues that, short of a comprehensive and undesirably violent revolution in the capitalist heartlands, the only solution to these problems is to combine much more powerful regulatory mechanisms by states for all their citizens with an agenda that deprioritises growth in favour of social and environmental agendas that are designed specifically to be equitable in the interests of the poor and the planet.

On definitions

Absolute and relative poverty

Definitions matter because they provide an indication of what people mean when they write or say words. All too often in the field of digital tech and development, terms are used imprecisely and vaguely, in part, deliberately to obfuscate, in part, to align with a particular hegemonic agenda, and, in part, through ignorance. Sometimes traditional meanings of terms are deliberately manipulated to serve a new ideological impulse. The term 'development' has thus evolved in meaning significantly over time, but it has also almost always retained positive connotations around progress for the common good (Unwin, 1994; 2017a). The 'development industry' that emerged from the mid-1940s alongside the political independence of states across Africa, Asia, the Caribbean, and South America has been widely reviewed, generating a plethora of differing opinions across all swathes of the political spectrum (see, for example, Rostow, 1960; Easterly, 2006; Stiglitz, 2006; Dercon, 2023). I use the term 'development industry' here to refer to the hundreds of billion £/¥/$/€ sector that include those from civil society organisations, consultancy companies, academics, UN agencies, bilaterial and multilateral donors, and foundations, as well as those who monitor and evaluate the interventions that they deliver (Mosse, 2011).

There is, though, a dissonance between much of the critical academic literature and the certainty of existing practice in international development. One of the main leitmotifs currently running through much of the academic literature on development is a 'post-colonial' agenda (Noxolo, 2016; Martinsone, 2024), which radically challenges the economic growth agenda outlined above and broadly advocates that the former colonial powers should pay retribution for their past exploitation in the expectation that this will enable economically poorer countries to rebalance their prosperity (although for a contrasting view, see Niemitz, 2024). As noted at the beginning of this chapter, for those in the world of practice rather than theory, the 'development industry' has nevertheless increasingly been concerned with eliminating poverty through enhancing the conditions for the private sector to deliver economic growth (see, for example, Sachs, 2005). In part, this continues to reflect the dominant power of economists within the sector (especially from the USA and Europe), but it is also increasingly evident that digital tech companies have become ever more prominent within this agenda through their role in expanding the global economy. Estimates in the mid-2020s (World Bank, 2024a) thus suggest that the ICT sector contributed some 5% to global GDP; at a national scale, the figure is often very much higher, with it, for example, being claimed that the digital economy represented 12% of India's GDP in 2024 (*Business Standard*, 2024).

In contrast to these approaches, this book adopts a different stance, focusing especially on the notion of development as the reduction of inequalities. This is based on the concept of *relative poverty* (see also Stiglitz, 2012), which is much more concerned with social, cultural, and ethical issues, and argues from a normative stance that development should be about the reduction of inequalities rather than the maximisation of economic growth. Those aligned with this approach often draw on Sen's (1999) work within the UN system, shifting the focus more towards 'human development', expanding human functionings and capabilities, rather than primarily focusing on economic growth (Sen, 2010). Sen's work has nevertheless also been criticised for its illiberalism, its under-theorisation of justice, its excessive individualism, and the difficulties of identifying appropriate information on capabilities (Wells, 2011). My own concerns with it are that it remains grounded in there being a sound economic basis upon which human development can then flourish, and that it insufficiently challenges the underlying structures in society that influence which functionings and capabilities people may have. I am more attuned with those radical alternatives that have suggested instead that the view of poverty as being absolute that underlies economic growth models needs to be comprehensively replaced by a relativist understanding of poverty that places emphasis on the social, political, and communal constructions of inequality (O'Boyle, 1999; Unwin, 2007). This, though, only applies if we want to use digital tech to create a better and fairer world for all, including the poorest and most marginalised. Disagreement with such an aspiration is indicative of the scale of the problem to be addressed.

Marginalisation, inequality, equity, empowerment, and emancipation

This book has already used terms such as marginalisation and equity that are all too often taken for granted and left undefined in works on digital tech and development. It is nevertheless important for the arguments that follow to be clear about the precise ways in which they are used here (for a more detailed discussion, see Unwin, 2017a).

The term *marginalisation* is used here primarily to refer to the processes whereby certain individuals and communities, such as people with disabilities, ethnic minorities, women in patriarchal societies, out-of-school youth and refugees, are deliberately, or subconsciously, prevented from fully participating in, and benefiting from, the societies in which they live (Graham, 2019; Ahmed, 2022; Varghese and Kumar, 2022). Concentrating on the processes (a verb) rather than treating these people as a product, the marginalised (as a noun) is for me more important because being aware of these processes promotes a better understanding of what needs to be done to change the conditions underlying them. Groups and individuals are not just 'marginalised'; we need to understand the processes of marginalisation that cause them to be so in the contexts where they are living.

Inequalities are the differences between people or places that are often considered to be unfair, and can be of many different kinds, such as social, financial, participatory, in terms of access to justice, and with respect to life expectancy.

They represent the opposite of equality. The normative judgement of fairness turns differences into inequalities. Given the dominance of the 'economic' in global discourses on development, it is not surprising that inequalities are most often considered and mapped in terms of economic criteria, using measures such as the Gini coefficient (Hasell, 2023), or the share of consumption or income held by different groups of people (see also World Bank, n.d.). It must also be recognised that inequalities vary at different scales and over time. It is thus possible to have a group of villages or cities with highly unequal societies in a country that overall may be relatively equal.

Equality and equity are often contrasted, with the word *equity* usually being used to refer to the fairness or justice in how people are treated. Whereas equality is about providing the same to everyone, equity recognises that people start with different circumstances and that more support needs to be given to those with the least so as to ensure fairness. As I suggest above, though, this neat distinction is not unproblematic, because without some consideration of ethics, inequalities are mere differences. These matters are of great importance with reference to access to digital tech, and especially infrastructure. Providing access to digital connectivity in isolated rural areas will usually be much more expensive than connecting everyone in a city with fibre networks. For equity to be achieved, the cost of that difference must be paid, and I argue that this is one of the most important responsibilities of government rather than the private sector, for whom such connectivity is usually unprofitable.

Empowerment and *emancipation* are likewise two rather different terms that are frequently used interchangeably, but have fundamentally different meanings. '*Empowerment*' is generally used with respect to digital tech in the sense of being a positive outcome that such technologies can have on development, but its full significance is rarely articulated in detail (Unwin, 2017a; Seth, 2022; Herzog and Hertwig, 2025). Literally speaking, empowerment means giving power to those without it, although the precise amount of power given will vary, depending on the socio-political system in place at a given time. Those in power who liberally speak or write about empowering the poor through digital tech, usually fail to realise that it actually implies a transfer of some of their own power to others. Power is not a limitless commodity that can simply be added to. Inglis (1997, p. 4) usefully summarises the distinction between empowerment and *emancipation* in the following way: 'empowerment involves people developing capabilities to act successfully within the existing system and structures of power, while emancipation concerns critically analysing, resisting and challenging structures of power'. This book grapples throughout with the fundamental question of whether digital technologies can indeed be used to empower marginalised people and communities, or whether this can only achieved through more radical emancipatory action. In one sense, we all need to be emancipated, freed from our illusions (Gross, 2010; Habermas, 1974; 1978); self-reflection by scientists and leaders of international agencies and companies working in the sector is one of the most important preliminary requirements for a fairer distribution of the benefits of digital tech to be achieved. If digital tech is indeed being used profoundly to enslave people, then we need to follow

Horkheimer's (1972) stance of a Critical Theory that seeks actively to 'emancipate people from the slavery of domination and oppression' (Unwin, 2017a, p. 28) and work to create a new society in which technologies are designed and used explicitly to serve the interests of all, rather than maintaining the power of a few Digital Barons, and the states where they are headquartered (see also Fuchs, 2023; 2024).

Finally in this section, something needs to be said about *poverty* and *the poor*. I recognise fully that someone, or somewhere, can be economically poor, but rich in culture. Indeed, I have argued strongly that Africa should be seen as a rich rather than a poor continent, and that Africans should build on that richness rather than listen to the global rhetoric that they are poor (Unwin, 2008). However, poverty is most frequently defined and described primarily in economic terms, and this is therefore the sense in which the word is mainly used in this book. Sometimes, I specifically use the term 'economically poor' for emphasis, but where this is not the case, the reader should generally assume that this is what is meant simply by the word 'poor'.

Rowe's vignette (p. 23) provides an opening challenge to set the scene for the remainder of this chapter. The story it tells can be replicated across most areas of digital tech interventions in international development. It nevertheless also serves as a reminder of the critical importance both of the harms that digital tech can be used to cause, as well as the lasting inequalities perpetuated by gender-based violence.

'The tech will save her': False promises in digital gender-based violence (GBV) solutions

Benita Rowe

(Designing digital strategies for service delivery in crisis and conflict settings)

Digital interventions aimed at preventing or responding to gender-based violence (GBV) have re-emerged in recent years in cyclical form, each positioned as a world-first innovation. Despite differences in format, these interventions often replicate a flawed set of design assumptions that consistently fail to account for lived realities.

Terms such as 'Open Source', 'secure', and 'safe' are often used to attract funding, yet usually describe tools that do not offer the protections these terms imply, nor implement the capabilities they publicly claim. A recent 'privacy-conscious' GBV-related digital platform deployed across several countries in the Middle East and North Africa (MENA) region, for example, received multi-million-dollar funding from several donors but was built on a general-purpose content management system, which was not designed for deployment in contexts where digital infrastructure must account for surveillance. While implementers described applying custom engineering to improve safety, the underlying architecture could not be reconfigured to address these risks. Furthermore, their plans to implement a moderated user forum as part of their drive to scale up introduced additional vulnerabilities. Governments often monitor such platforms, and implementers cannot detect or prevent such surveillance, yet they continue to assure users that their platform is safe and private.

These issues are not limited to digital platforms. Although packaged differently, other interventions rely on the same flawed assumptions about infrastructure, access, and responsiveness. A widely promoted wearable device developed to support women's safety by discreetly triggering alerts during an attack is another such example. The device connects to nearby mobile phones via Bluetooth and uses GSM networks to notify responders. However, in many high-risk, low-connectivity environments where these devices were meant to be deployed, Bluetooth pairing is unreliable in moments of distress, mobile signals are weak or deliberately disabled, and responders are often out of range or unable to act in time.

These constraints are well documented, but these two examples (of many) illustrate how initiatives continue to be promoted as innovative solutions for populations they are structurally incapable of serving. Too often, tools are optimised for the visibility of their implementers rather than for the realities of the people they are funded to support. When the intended users are GBV survivors, there is frequently an implicit framing that positions them as the problem to be solved through technology, rather than addressing the structural conditions that cause or prolong harm.

Increasing inequalities: the role of digital tech

How and why digital tech is created and used to increase inequality

Most digital technologies are conceived, designed, created, and used, either deliberately or with insufficient thought, by individuals for their own (or their organisation's) benefit, and thus increase inequalities. It is crucial to recognise that it is not the technologies themselves that do this, but rather it is the *interests* of those who produce them that are the ultimate cause. The concentration of wealth in the hands of the super-rich is increasing significantly, with much of this wealth being generated by and through digital tech. Oxfam (2024) thus estimates that the richest 1% in the world have more wealth than the bottom 95%, and this wealth is becoming more concentrated. Moreover, eight of the ten wealthiest people in the world, and all the top four, in November 2024 had gained their wealth through digital tech (*Forbes*, 2024a; see also Joseph, 2022 to highlight the wider economic dominance of the major US tech companies).

Technological developments are not in any sense natural or autonomous (Feenberg, 1991), but instead they are designed and created by people with particular interests to serve purposes that are almost always intended for their own benefit in some way (Unwin, 2017a). Language is again very important here, because it shapes our understandings and therefore our ability to change things. It is critically important for those in the field to shift away from using phrases implying that digital technologies cause anything themselves so that the misleading notions of *technological determinism* (the belief that technology has some causal autonomy separate from human design) and *instrumentalism* (the belief that technology is a neutral instrument) can be avoided (Veblen, 1921). We must therefore move away from statements such as 'Digital technologies are dramatically transforming our world', as in the opening words of the Global Digital Compact (UNOSET, 2024, p. 1), and replace them with phrases such as 'Those who create and sell digital technologies are dramatically transforming our world'. This shift in language has enormous ramifications for how we can effect practical change on the ground, and goes beyond many of the widely used arguments associated with research on the social shaping of technology (Dutton, 2013) by placing emphasis primarily on the central role, and thus responsibilities, of individuals in driving inequalities and harms resulting from their use. It is these individuals who need to be persuaded to change what they are doing if the world's poorest and most marginalised are indeed to benefit.

In essence, companies within the present capitalist system have two fundamental needs: (1) to increase their market (either size or share), and (2) to reduce costs (traditionally mainly the cost of labour power). Digital technologies have been shaped to be remarkably adept at serving both of these needs, enabling the world to be connected almost instantaneously and dramatically increasing labour productivity in many areas of employment. Equally importantly, they are being designed to do so at a rapidly increasing pace, thus accelerating the turnover of capital. The potential for the use of AI and robotics to transform the relationships between labour and capital is considered in more detail in Chapter 6, but for the moment it is important to recognise that the structures that have shaped them rely on inequalities, primarily so that companies can always find those willing to sell their labour power more cheaply than the cost of their existing employees. The use of call centres in different parts of the world and part-time labour working from home, both enabled through the use of the Internet, are classic examples of this reliance. Companies also need ever-increasing markets to overcome the tendency for their rates of profit to fall (Moran, 2020).

The use of digital technologies dramatically increases inequalities and also causes new harms. Chapter 1 illustrated this clearly in the example of the ways through which a rich person simply buying a new advanced mobile device actually makes the world more unequal. Any use of digital tech, though, will drive the process of marginalisation if it is beneficial to some, but not to all. Those who cannot afford the technology, those who cannot access it, those who do not know how to use it, or those who are prevented from using it, will all become further disadvantaged unless mitigating actions are taken. They are also more likely to suffer from the many harms caused by the use of digital tech (Bartlett, 2015; see also Rowe, p. 23), such as online scams targeting the elderly (Age UK, 2023), online grooming (Wefers *et al.*, 2024), or sexting and child pornography (Internet Watch Foundation, 2023). To be sure, those who are above a notional basic absolute poverty line and can afford devices and connectivity may well be able to gain some benefits from digital tech, but, even then, only if they can use it safely, wisely, and securely. The poorest and most marginalised are nevertheless becoming daily more marginalised through the use of digital technologies by those richer and more powerful than they are. It is salutary to recall that a third of the world's population was not using the Internet in 2023 (ITU, 2023). The notion of trickle-down or filtering-down economic growth benefitting the poorest has long been disproven (Stiglitz, 2016), but still persists in the minds of many advocates of the rhetoric that the use of digital tech will eventually benefit all people, and that all we have to do is connect them to the Internet. It must never be forgotten that the propagation of digital technologies exacerbates differences and is a driver of inequality. It is simply not possible to use a driver of inequality to reduce inequality, and thus reduce poverty or empower people when defined in relative terms.

This tech-enabled inequality can be illustrated by numerous examples across all sectors of the economy and society. Fintech is thus often seen as being one of the main ways through which poverty can be reduced and the SDGs delivered through its beneficial use in expanding financial inclusion, increasing access to financial services, and improving the effectiveness and efficiency of aid delivery systems (Mhlanga, 2023). However, it is also a way by which those building new financial systems have sought to increase their markets by incorporating poorer people with the intention of exploiting and manipulating them for profit. These systems also increase vulnerability, and although rich people can often afford to be scammed, even the loss of a small amount of money from digital systems can be life-changingly devastating for poor people (Gupta, 2024; Keshvani-Ham and Tiwari, 2025). In Kenya, where mobile transactions account for more than half of the country's GDP and Safaricom controls 99% of the market through its M-PESA service, 47.4% of customers surveyed were reported to have experienced fraud in 2021, often through SIM-swap scams (KNBS and FSD, 2021; see also Siele, 2023).

Edtech is likewise usually seen as providing significant learning benefits for the world's poorest children, as evidenced by the creation of entities such as the EdTech Hub, funded by the UK International Development, the Bill and Melinda Gates Foundation, the World Bank Group, and UNICEF, with the intention of providing evidence of how such interventions can be used to enhance learning outcomes (see vignette by Hollow, p. 126). During much of the 2010s, mobile devices were seen as providing very positive hope for improving learning for young people, with UNESCO convening its annual Mobile Learning Week between 2011 and 2022, with its name subsequently being changed to Digital Learning Week in 2023 (see also West, 2023). Yet, increasing numbers of countries are now considering banning mobile devices in schools because of their perceived harms not only for learning but also for the mental health of children. Companies building devices, designing social media apps, and providing learning software all depend on an increasing market to build their profits, and are enabled to do so by capturing large numbers of young people more or less untrammelled until the mid-2020s (for an alternative positive model of digital engagement with young people, see vignette by Longmore, p. 27). It is only recently that some governments have begun to question this, with Australia announcing in 2024 that it plans to ban social media for children under the age of 16. Moreover, for digital technologies to be used in schools, there also has to be electricity to power them. While most people in India and China have potential access to grid electricity, only 57% of people in Africa did so in 2022 (IEA, 2022). Those without reliable electricity cannot benefit from the potential educational or health benefits of using digital tech, and so become further marginalised.

The positive advantages of using digital technologies, especially mobiles, for health are also very widely promoted. The World Health Organization thus comments:

> Digital technologies are now integral to daily life, and the world's population has never been more interconnected. Innovation, particularly in the digital sphere, is happening at unprecedented scale. Even so, its application to improve the health of populations remains largely untapped, and there is immense scope for use of digital health solutions.
>
> WHO is harnessing the power of digital technologies and health innovation to accelerate global attainment of health and well-being.
>
> (WHO, 2024)

Such promotional statements are a boon to the ears of those companies providing digital health apps, as well as those building devices and the infrastructure to enable their use. There are certainly many benefits to be gained from the use of such technologies, but this only applies to those who can afford, access, and know how to use them safely. There is much evidence to suggest that the elderly, those who speak or read minority languages, and those who cannot afford or access the recommended treatments become further marginalised as a result. Moreover, there are fundamental questions to be asked about health data security and privacy, especially when patient systems are sold to the private sector or personal data hacked (BMJ, 2024; for recent examples in the UK, see Francis, 2024; Macon-Cooney *et al.*, 2024).

The youth-led imperative

Janet Longmore

(Visionary CEO transforming youth as tomorrow's digital leaders of change)

We've seen it all before – for decades. Youth are the majority age group in developing countries, disproportionately unemployed in marginalized and disadvantaged communities, and yet, the energy, innovation and creativity of young women and men are a wasted asset – a drain, and not a driver, of economic and social change. *We've seen it all before* – youth voices demanding to be heard, impatient to contribute to their families, peers and communities, to drive their futures – assertively acting out their frustrations. And, of course, *we've seen it all before* – half-hearted lip service to youth engagement, token participation on committees and corporate boards. They are there to be seen rather than to be heard – a 'tick' in an inclusion box. They are usually patronized spectators in adult-led decision-making.

Must we see it again? The answer is an emphatic "NO!" For over 20 years, Digital Opportunity Trust (DOT) has been providing young people across the world with essential innovation, entrepreneurship, digital and leadership skills, and the opportunities to apply them, in the belief that by transforming the individual, the individual can transform the community, investing in young people as agents of change. The results are compelling. Ariane, of Rwanda, founded Solar Light to provide affordable solar energy to communities that lack reliable electricity. Olivier, of the DRC, founded the Refugee Innovation Centre in Uganda which delivers digital skills and entrepreneurship training to hundreds of his peers. Young leaders such as these drive change, inspire their peers, network with each other, form teams to fill service gaps, and operate at the intersection between the potential of youth and the use of digital technology.

DOT's (2022; 2023) experiences suggest that six key recommendations need to be incorporated into initiatives seeking to inspire and support young people through the use of digital tech (see also Figure 7.2 in Chapter 7):

1. *Listen to them.* Engage in meaningful conversation where they are, and reflect on what you will learn from the front lines of community reality.
2. *Peers and partners.* Abandon hierarchical tendencies and recognize and respect young women and men as peers. Bring them into your formal governance instruments, committees and boards.
3. *Co-design, co-create and co-deliver.* Peer-to-peer learning is effective and efficient, and peer-to-peer projects have better outcomes than do hierarchical models. Therefore, introduce hybrid processes in which youth co-participate in the design, development and implementation of policies, procedures, systems and products.
4. *Empower.* Provide transformative skills and practical opportunities for youth to apply them in digital work, entrepreneurship, social enterprise, and leadership.
5. *Resist yourself.* Resist the temptations to interfere, maintain a light touch, do not stifle the creativity and innovation of youth with conventional experience and wisdom. Let it be invited.
6. *And trust.* Trust is the basis for effective collaboration. Mutual trust strengthens self-esteem and reinforces the self-confidence necessary for young people to apply their new skills.

Applying these six 'learnings' provides a fundamental 'mindset shift' that reflects new skills and an entrepreneurial socially responsible spirit, resiliency and adaptability among young people that are critical for navigating employment and self-employment opportunities in a digital economy.

Why using terms such as 'bridging the digital divide' is so harmful

The term 'closing' or 'bridging' the 'digital divide' has long been used to describe how digital tech can be used to create a better world (Servon, 2002; Milward-Oliver, 2005). The Institute of Electrical and Electronics Engineers (IEEE, 2024), for example, positively addresses this issue of solving the digital divide in terms of creating 'a more equitable future'. However, its recipe for doing so is largely based on technological solutions: increasing connectivity, better digital literacy, providing affordable and equitable access, investing in digital education, devising locally appropriate public-private solutions, and developing stronger broadband infrastructure. It is far from alone in using such arguments. Yet digital tech itself cannot solve the problems caused by the design and use of digital tech. It is not a 'digital divide' that is the problem. Digital technologies did not create the existing structures of global inequality, but their use has dramatically increased these inequalities. Technological solutions cannot solve a non-technical problem. The poor are now relatively much poorer than they were at the start of this millennium, in large part as a result of the ways that digital technologies have been designed and promulgated. It is misleading to consider these inequalities as a divide that can be bridged, because the design and use of digital tech are actually making the gap bigger.

Continued use of the notion of 'closing/bridging the digital divide' is so harmful because it perpetrates the myth that it is possible to use technology to solve a problem that is actually made worse by the use of digital tech. It also feeds into a scientistic view of the world, and reinforces instrumentalist and deterministic views of digital tech. Providing enhanced connectivity, digital literacy, affordable and equitable access, digital education, public-private solutions, and broadband infrastructure will not automatically enable the world's poorest and most marginalised to catch up with the benefits that the richest and most powerful gain from the use of such technologies. It is no coincidence that the rise in inequalities within countries referred to in the World Inequality Lab report and noted in Chapter 1 (Chance *et al.*, 2022, p. 11) closely coincided with the rapid expansion of digital technologies. While inequalities between countries have declined over the last 20 years, those within countries have increased dramatically. As the authors note, 'This sharp rise in within country inequalities has meant that despite economic catch-up and strong growth in the emerging countries, the world remains particularly unequal today' (Chance *et al.*, 2022, p. 11).

Continuing to use a rhetoric that it is indeed possible to bridge the digital divide is also harmful because it gives hope to the poor and marginalised that the rest of the world might actually do something about it. In Africa, for example, it gave rise to notions such as 'cheetah pole vaulting' coined in 2005 by the Ghanaian Ekwow Spio-Garbrah, reflecting his belief that Africans could run as fast as a cheetah and jump as high as a pole-vaulter to overcome the digital divide (The Bee, 2005). Despite ambitious programmes across the continent, commitments to a so-called Fourth Industrial Revolution in Uganda and South Africa, the creation of a Silicon Savannah in Kenya, and innovation hubs across the continent over the last 20 years (see also Chapter 5), the reality is that Africa has not yet leap-frogged other parts of the world. Most of its digital systems remain relatively further behind those of most other parts of the world (Bhorat *et al*, 2023). There is, nevertheless, hope that these challenges can be overcome, and Ileleji's vignette (p.30) highlights one such more positive perspective. This particularly emphasises the needs of grassroots organisation to deliver relevant interventions.

A Gambian perspective on digital tech and equity

Poncelet Ileleji

(Digital Rights activist, championing the African Union Agenda 2063)

The beauty of digital technology initiatives is the socio-economic impact they can have globally. The growth of Afrobeats and the music industry in Nigeria is a classic example of this, building on the talents of young Nigerians in producing music but enabled by two disruptive digital technologies: first, Google's YouTube helped them reach millions; and, second, the establishment of Internet Exchange Points in Nigeria and other African countries enabled Google Cache proxies to be localized outside America and Europe.

Progressive digital disruptions must also be based on equity and a grassroots emphasis to ensure inclusiveness. For example, the use of citizen-generated data can help provide an inclusive approach to addressing inequalities, especially when it is combined with locally grounded innovation hubs, such as Jokkolabs in Gambia which creates an organic community of digital entrepreneurs working with the local community to help solve their problems.

For initiatives, such as the UN's Global Digital Compact to succeed, especially in sub-Saharan Africa where only about 30% of the people are connected to mobile Internet, citizens must be involved in the multistakeholder approach at the grassroots level. Allowing communities to be engaged through data-driven approaches makes it possible to have holistic and sustainable programmes where digital technologies can be used meaningfully within the value chain of provision of services, as in the rise of Afrobeats. The collective need for organisations starting at the grassroot level to invest in community networks can help citizens engage better and adopt digital technologies that can improve their lives, be it in innovation, entrepreneurship, health or education. In the end, you build digital sovereignty starting at the community level and create an ownership model that involves communities from the very beginning.

We can learn important lessons from the example of the rise of Afrobeat's globally. Its inclusive approach, based on equity for all irrespective of where they live, has enabled significant change. Initiatives such as the Global Digital Compact must likewise start at the grassroots level that embeds equity for all globally. International civil society organisations, such as the Association for Progressive Communications (APC) have a key role to play in this, not least by sharing good practices of what works, and by advocating for the importance of civil society in delivering appropriate Internet-based applications at the community level.

Why the link between digital tech and economic growth is so powerful

The twin mantras of economic and demographic growth

The dominant capitalist global economic system is fuelled by the twin mantras of economic growth and demographic growth. Both are essential for its survival, primarily because of capital's continuing need for ever-expanding markets and labour inequalities (Unwin, 2017b; 2023b). This has many consequences, not least their significant impacts on the environment that have long been recognised (Meadows *et al.*, 1972; Colman, 2021). As noted in Chapter 1, digital technologies have been a key component of economic growth over the last 20 years, and are forecast to become even more significant through the use of AI and quantum computing in the future. It is almost as if the interconnected digital system founded on the Web and the Internet has become a black hole in space-time from which nothing can escape.

However, although digital tech is widely used across the world, value added in the two main industry sectors of ICT manufacturing and ICT services is highly concentrated. China and India account for more than half of the global value added in these two industries, and the World Bank (2024a, p. 26) reports:

> the top six economies account for 80 percent of global value added in ICT manufacturing and 70 percent in ICT services. These shares have increased further since 2010, as economies of scale and scope, network effects, and winner-takes-most characteristics of the ICT sector have cemented and escalated dominance by the world's leading economies.

Again, at the macro-scale, this is a further expression of the global inequalities that shape, and in turn have been enhanced through, the production of digital tech. Nevertheless, the fundamental changes in, and rapid growth of the BPO-ITES (Business Process Outsourcing-Information Technology Enabled Services) sector brought about through digital tech do theoretically provide some opportunities for poorer countries to create new jobs that the World Bank (2024a) sees as combatting the brain drain.

Such economic growth has very significant material and environmental costs, and insufficient attention has yet been paid to developing a holistic understanding of the widespread and increasing environmental harms caused through the creation and use of digital tech (DESC, 2021). However, more generally, there is now a growing body of literature challenging the basic assumption that GDP needs to keep on rising as fast as possible. As Hickel (2020, p. 165) has argued, 'people are willing to go to extraordinary lengths to justify the continued pursuit of economic growth', and 'This narrative relies heavily on the claim that technology will save us' (Hickel, 2020, p. 127). D'Alisa *et al.* (2015) likewise reject the illusion of growth and advocate a repoliticisation of the public debate colonised by the idiom of economism that gives primacy to economic goals and interests. Although they do not focus at any length specifically on the role of digital tech, their questioning of the need for growth is very important, and needs to be taken seriously by those continuing to promote growth through digital tech. A combination of a critique of both scientism and economism that underlies the arguments in the present book offers real potential to reconfigure the use of digital tech away from the growing inequalities and harms that it causes, and towards greater equity through increased emphasis on social, cultural, and political interests of the many rather than the few (see also Colman, 2021). It is time to shift the balance away from digital tech being used to promote growth, to its design and use explicitly to achieve equity.

We are the data – and seem to like it: modern indenture

One of the most significant digital developments in recent years has been the rise in data analytics, enabled in part through cloud computing. Fortune Business Insights (2024) thus estimates that the data analytics market size, dominated by the USA, was worth US$ 41.05 billion in 2022 and is expected to rise to US$ 279.31 billion by 2030. Many see this as a central element of a so-called Fourth Industrial Revolution (Schwab, 2017; for a critique, see Unwin, 2019), but the idea of profiting from data was born during the 1990s in the early days of the Internet. Finding a way to profit from traffic on the Internet was rapidly solved by companies such as Amazon (originally founded in 1994 as Cadabra) and Google (created in 1998), and was subsequently turned into a fine art by the likes of Facebook (founded in 2004, now part of Meta), Twitter (in 2006), 微博 (Weibo, in 2009) and 微信 (WeChat, in 2011). Combining multiple data sets has enabled vast amounts of data about individuals to be combined, not only for purposes of marketing but also for surveillance. As Cheney-Lippold (2017) titled his book, *We Are the Data*. To be sure, data analytics has also become a powerful new tool to try to improve decision making by governments and companies, and has been seen once more as an important engine of economic growth and delivering the SDGs. However, it ultimately comes down to a question of power: who has the ability to use these data? Unless the poor and marginalised are able to use data analytics themselves to serve their own interests, they will remain at the behest of, or under the control of, the rich and powerful who are able to use these technologies primarily in their own interests.

The key trick has been to get people to provide their valuable data without the companies having to pay for it. This has been achieved primarily by offering them a 'free' service that it was cheaper to provide than the ultimate value of the data that companies acquired. For this to work, users had to be kept in the dark about the real value of their data. This process has been highly attuned so that it can be seen in many ways as a new mode of production. Companies are able to extract surplus profit from the value created by users, and to make it more effective much of this is done in users' 'leisure time', or when 'working' for another company as a form of 'proxy labour'. This is, in a sense, a modern form of indentured labour, whereby users contract themselves and their children as labour to someone else without a salary for a period of time. Is this so very different from letting children be exploited through the Internet? Signing the terms and conditions associated with these apps is not dissimilar to the agreements made by indentured labourers in the nineteenth century. Furthermore, when people cannot stop using their digital devices, they have then lost their freedom and should instead be seen as having become enslaved by the Digital Barons who control these companies. To make matters more shocking, they are even persuaded that they are actually enjoying their indentured or enslaved lives, without realising the harms that it is causing them. The greater the number of poor people who join and use these platforms, the richer the Digital Barons become; inequalities inexorably increase (see Chapter 6 for the development of this argument).

Additionally, the companies developing digital tech have also been at the forefront of arguing that governments should provide all their information about citizens for free as 'open data', even cajoling academics into their cause (see Chapter 6 on openness). Academic research on health data provided freely to them by governments, for example, contributes extensively to the profits of pharmaceutical companies who develop and sell new vaccines or treatments based on their research (for a critique, see vignette by Juliette Unwin, p. 85). Likewise, data extracted by satellite remote sensing from farmers' use of land is used to enhance agribusiness profits without the farmers necessarily seeing any direct benefits from their labours.

Towards an equity agenda

The arguments of this chapter are based on two fundamental assumptions:

- The moral case that existing global inequalities are wrong, and should be reduced.
- A belief that digital technologies can indeed be designed and used to create greater equity in the world.

Not everyone will agree with these. Some do not really worry about inequalities and will want to exploit them to continue to maximise their profits and income (see Rawls, 1971; 2001; O'Neill,1996; 2016, for arguments as to why they are wrong in so doing). Some will not accept that digital technologies can ever be used to reduce inequalities. Some will defend their position that increasing economic growth is the only way to reduce poverty. Others will argue that the multistakeholder model being used today is delivering on poverty reduction very well, thank you, and just needs more money to be ploughed into digital tech so as to deliver the SDGs. Hopefully, the arguments in this chapter may encourage all of us to think again, better to understand the issues and question our own taken-for-granted assumptions about the intersection between digital technologies and development. This is a first stage to enlightenment about digital tech.

Both and . . .?

There is growing recognition that the SDG agenda is imperfect, and that digital technologies may not have been used most effectively in achieving even these goals, let alone a truly fairer and better world for all. This may have been one of the more positive reasons why UN Secretary-General Guterres and those in his Secretariat, sought to include a Global Digital Compact in the 2024 Summit of the Future. Back in 2020, at the height of COVID-19, Guterres acknowledged that the pandemic 'had demonstrated the fragility of our world . . . revealing fractures in the fragile skeleton of the societies we have built' (Guterres, 2020a). He went on to exemplify and assert simply and clearly that 'Inequality defines our time'. He even stated specifically that digital transformation could widen inequalities even further. Why has this call not yet been sufficiently addressed?

Unfortunately, Guterres (2020a) went on to say, 'The Roadmap for Digital Cooperation, launched at the United Nations [the previous month] promotes a vision of an inclusive, sustainable digital future by connecting the remaining four billion people to the Internet by 2030' and that 'Technology can turbocharge the recovery from COVID-19 and the achievement of the Sustainable Development Goals' (see Chapter 3 for a more detailed discussion). Sadly, more of the same is absolutely not the answer. Technology is not, as he implied, an 'agent' that is somehow separate from those who conceive and design it. It is these people who must make choices. They can continue to seek to maximise profit through economic growth, combining economism with scientism, or they can choose to develop digital tech that is explicitly designed to reduce harms, to minimise environmental impact, and to reduce inequalities. All digital tech, for example, could be designed so that as far as possible it maximises its utility for those with disabilities, rather than requiring them to purchase additional assistive tech in order to be able to benefit from it (see vignette by Gardezi, p. 34). Great steps have been made in this direction in recent years, as with Apple's and Microsoft's operating systems' accessibility features, but much more remains to be done and people with greater disabilities generally still have to pay more for their use of digital tech than do those of us with fewer disabilities (Eid, 2017). Is this right, or fair?

Why persons with disabilities will keep on experiencing failure

Akber Gardezi

(Honest, compassionate, empathetic, thoughtful, supportive, inclusive, pragmatic, curious, scholarly, collaborative)

Much of my work over the last decade has focused on initiatives related to technology for Persons with Disabilities, mostly in the Islamic world. The over-riding conclusion from these experiences is that most initiatives promoting digital assistive technologies become stuck in a repetitive cycle of initial enthusiasm, ambitious launches, and subsequent stagnation or abandonment. Why is something so important, so unsuccessful?

First, there remains a fundamental gap in awareness and understanding of digital assistive technologies. In much of the Islamic world persistent stigma and misconceptions regarding disability limit the adoption and effective use of digital interventions. Most communities lack sufficient knowledge or encouragement to use digital tools that could significantly enhance individuals' independence and societal participation.

Second, digital assistive technology projects commonly lack cohesive, sustainable frameworks. Efforts often prioritise short-term visibility over long-term infrastructure and support, leading to poorly maintained or abandoned digital interventions. Initial investments are often not accompanied by training, support, and updates, which diminishes their effectiveness and longevity.

Funding challenges further compound these issues, through a cyclic dependence, primarily on Western donors. Such funding comes with rigorous audits and scrutiny, but it is rigged to facilitate the corrupt. For those not on the take, even well-intentioned expenditures can be viewed suspiciously, discouraging local innovators and causing genuine change-makers to lose enthusiasm and confidence.

It is also common for incumbent governments to launch digital initiatives, knowing very well that they require robust Internet connectivity, widespread digital literacy, and compatible devices. However, many regions lack the essential digital infrastructure necessary to support effective and accessible assistive technologies. Essentially these initiatives turn into political parlour tricks, that reveal no real intention to support the most marginalised.

Furthermore, civil society organisations involved in digital assistive technology initiatives often suffer from competitive rather than collaborative environments. Petty organisational rivalries and dynastic leadership approaches prevent coordinated advocacy and efficient resource utilisation.

Figure 2.1 Technology for the visually impaired: (a) conventional Braille frames at an Institute of Education in Tunisia, and (b) a digital magnifier at the Azad Kashmir Association for the Blind School

Source: Tim Unwin.

Persons with disabilities time and time again voice their concerns that effective digital assistive technology initiatives require active participation from the community. However, these projects are too often developed without meaningful consultation with the intended users. Instead, a handful of cherry-picked token Persons with Disabilities are selected, based on the whims of funding organisations. The resulting solutions are branded as localised, indigenised and on-the-ground innovations, but in reality they are seldom aligned with real needs.

Realistically, it is difficult to imagine that the existing global development growth agenda will be abandoned soon. There are too many powerful industrialists, economists, international agencies, and scientists with vested interests in its maintenance to permit that (see also Chapter 4). It is, however, possible to imagine that the growth agenda could be balanced by an equity agenda, and to work much more explicitly to ensure that digital tech developments focus on both. Unfortunately, the outcome document from the Summit of the Future (the Pact for the Future; UN 2024b) only mentions 'inequality or inequalities' ten times and 'equity' five times in its 56 'Actions' over 36 pages; 'economy or economic' is mentioned 39 times (see also UNOSET, 2024). The economic growth agenda would seem to remain all powerful.

In practical terms, there are many things that individuals and organisations can do to shift the balance away from the 'digital for growth' agenda to 'digital for equity'. Some of the easiest to achieve are:

- Using language more precisely, as discussed above, especially to counter the rhetoric of technological determinism and instrumentalism.
- Ensuring that all new digital initiatives consider both their impact on equity as well as on growth and profits. This applies as much to the creation of new technologies, as it does to programmes of digitalisation implemented by governments or organisations in the interests of 'development'.
- Focusing more on the reality than the rhetoric, and being more realistic in claims made about digital tech (see also Chapter 3).
- Choosing to work in the interests of the poor and against the structures of marginalisation, rather than for the rich and powerful (see Chapter 4).
- Using cash rather than cards or digital IDs, maintaining multiple identities, and reducing online presence, especially on social media.
- Championing the right to be unconnected, more than to be connected (see Chapter 7).

Reforming regulation in the interests of the most marginalised

Governments at all scales have a fundamentally important role to perform in balancing the distribution of wealth in their polities, and to ensure that the weakest and most vulnerable are cared for, at least in recent definitions of civilised societies. This latter principle has been espoused in various forms from the eighteenth century onwards by people as diverse as Dostoyevsky, Darwin, and Jefferson. However, the rise of neoliberalism, especially through its practical implementation in the 1980s in the USA and much of Western Europe, with its emphasis on privatisation and deregulation to maximise economic growth and well-being, shifted the balance of power fundamentally away from governments and towards the private sector. This laid the bedrock upon which private telecommunication companies were able to innovate and begin dramatically to transform the digital tech sector in pursuit of unfettered profit (Gentzoglanis and Henten, 2010).

There is now, though, growing recognition in many parts of the world that increased regulation of the telecommunications and digital tech sectors is essential (Unwin, 2017a). This has been particularly strong in Europe since the early 2000s (Simmons and Simmons, 2022), and especially since the introduction of the General Data Protection Regulation (GDPR) in 2016 (to become effective in 2018) and the Digital Services Act in 2022 (European Commission, 2022). These are widely seen as being attempts to rebalance power between European citizens and digital tech companies, especially over data and harmful content and practices. Leaders of US companies have, in contrast, frequently been critical of such regulation, arguing that it reduces economic growth and innovation, but European politicians have been willing to accept this in the interests of their citizens' well-being. Paradoxically in the USA, with its championing of individual rights and embracing of economic and financial success, which have enabled its most important digital tech companies to flourish, once these companies get too successful, they then suffer from accusations of being too big and monopolistic, as with the 2023 and 2024 filings against Apple and Google for violating anti-trust laws and building monopolies (DOJ, 2023; 2024). In contrast to these polities, Chinese regulation has been much stricter, and it has long been a state where data privacy and sovereignty have been treated particularly seriously (Zhou *et al.*, 2024). China has thus placed strong emphasis on controlling data both for national security purposes and for individual privacy within the framework defined by the government (InCountry, 2024). Global differences between digital regulatory environments remain a key area of international discord, reflecting fundamentally different ideological positions and interests that have significant impacts on growth and equity, with the more regulated environments often tending towards greater equity, and the least regulated promoting greatest growth, but at the same time greater inequality (Perez, 2021; GSMA, 2024a).

On a more practical level, telecom regulation plays a very important role in influencing the affordability of access to digital tech. However, pricing and affordability also reflect the competitive environment and company policies. In India, for example, the aggressive pricing policies of Jio, which launched in 2016 with a free trial offer, grew within two years to be the third largest telecom operator in the country, and has played a significant role in lowering prices of connectivity across the country. Globally, North and South America tend to have the highest absolute mobile data pricing, and northern Africa and Asia the lowest, but there are some considerable variations within regions, and African prices generally remain stubbornly high, especially when considered as a percentage of incomes. The affordability of mobile data as a percentage of income in low-income countries is thus on average more than three times the figure in middle- and upper-income countries, thus perpetuating the inequalities between them (Rodriguez and Woodhouse, 2022; see also Fleck, 2023).

One, contested, way that some governments have tried to support access to infrastructure for their poorer citizens has been through the creation of Universal Service Funds (USF; also Universal Service and Access Funds, USAF), which require operators to pay a levy that is then used to subsidise infrastructure and equipment in marginal contexts. These were first established in the USA in 1997, and have since then been used widely in lower-income countries as a means purportedly to enable rural areas and poorer communities to benefit from digital tech (World Bank, 2005). While intuitively they may seem a reasonable way to try to reduce inequalities, many operators have argued that they could deliver these services more cheaply and effectively without such levies. Moreover, a comprehensive GSMA report (Ladcomm, 2013) has provided strong evidence that such funds were at that time inefficient and ineffective, with many only disbursing a small amount of the income they generated. More recently, however, government representatives from several African countries at a CTO meeting in London in July 2025 described in considerable detail the benefits that such funds brought to marginalised communities in their countries.

From rhetoric to reality

This evidence suggests that attempts at tightening regulation and enforcing companies to deliver on their obligations for universal access remain challenging, especially in many of the poorest countries (Unwin, 2017a). However, the need for tougher regulation of infrastructure, connectivity costs and content has been increasingly widely seen over the last decade, especially in Europe, as an essential means to constrain some of the more harmful effects and excessive profits of the digital tech sector. The contested nature of regulation, as companies seek to maintain their dominance over what is often seen as government interference was, for example, clearly witnessed in the open letter written by scientists against some of the proposals in the Global Digital Compact mentioned in Chapter 1 (Applequist *et al.*, 2024). To be sure, increased regulation of the sector is likely to reduce economic growth, but this only highlights that growth and equity are incompatible. Having more equity for less growth would seem to be a wise choice for those involved at the interface between digital and development who really want to reduce relative poverty. A first step towards such an outcome would be to change the rhetoric away from such problematic notions as 'bridging the digital divide' and to focus much more rigorously on changing the reality on the ground. It must never be forgotten that the apparently inexorable rise of digital technologies during the twenty-first century serves a particular set of interests that thrives on the necessity of inequality; it is in no way natural or inevitable. Recognising that this is only going to lead to greater poverty is a crucial first step in seeking to redesign digital technological systems that can help to achieve greater equity.

There will be those who see such a task as impossible, who argue that we must retain economic growth, and that such growth has already lifted countless people out of poverty. There will be many who even relish this opportunity to maximise their status and wealth through their work in the digital tech sector, arguing that the poor will always be with us, and that there is nothing wrong in accruing great individual material success. However, this has left 'those without' even further behind. Ultimately how we choose to react to the arguments of this chapter will depend on our varying moral compasses (explored further in Chapter 7), but whatever these are, it is time that we stopped pretending that digital technologies are a panacea for 'development', and will inevitably make the world a better place. Chapter 3 explores the implications of such arguments for policy making and practice at the global level through an analysis of the way that much of the UN system has been subverted by the interests of global digital capital.

All of us are 'a piece of the continent'; anyone's unnecessary death diminishes us all. It is time for all those engaged at the interface between digital tech and development who are truly concerned with the needs of the poorest and most marginalised to look deeply inside themselves and ask why we are allowing such inequalities not only to persist but actually to get wider.

The geopolitics of digital knowledge

Domenico Fiormonte

(Academic researcher, Universities of Roma Tre and Dar es Salaam)

> Every relationship of 'hegemony' is necessarily a pedagogical relationship and occurs not only within a nation, between the various forces of which the nation is composed, but in the international and world-wide field, between complexes of national and continental civilizations.
>
> (Gramsci, 1971, p. 350)

Nietzsche maintained in 1874 that 'historical knowledge' is fundamentally responsible for manipulating the past and hence puts the present both at risk and on hold (Nietzsche, 1980). His main argument was that *no analysis, recovery or reconstruction of the past can ever be a neutral process*. In so doing, he was referring to the perennial and continuous work of human interpretation: hermeneutics was at the core of history. But how do we make our interpretations today? How is the infinite thread of material posted online forming the basis of our present and future interpretations?

In the early 1990s I was a young 'digital philologist', trying to build digital archives of literary texts. I liked to think of my job as someone who was struggling to bring the past into the future. In creating online resources, I realised that *digital representations* need and imply different levels of codes, tools, and infrastructures. Each of these layers hides a cultural, linguistic, and epistemic bias. Technology, like history, was not a neutral process! After reading Nietzsche, I discovered Gramsci: not only are our tools intrinsically biased, but they have *owners*. Digital knowledge is therefore *geopolitics*.

When I started to encode my texts, I used ASCII (American Standard Code for Information Interchange). The most commonly used Web languages, such as HTML, are still based on the 126 symbols of the ASCII table. Later we moved to UNICODE (Universal Character Encoding), the system providing standards for potentially encoding all living (and much of the historical or even fictional) languages. UNICODE's Board of Directors is made up of eight to ten members, all coming from Big Tech companies, including Google, Facebook, Adobe, Apple, Airbnb, Microsoft, and Netflix. No public or educational institutions are represented. These are just examples, but they illustrate clearly the geopolitical structuring of the basic elements of the Internet.

So who has the power today to 'represent' digitally the world's languages, the core of human cultures? It is a group of Western, predominantly English-speaking and US-based corporations. The issue is even more evident when we look at the limited number of languages supported by most used social media and Web tools. According to data provided in 2020 by internetlanguages.org, WhatsApp supports 58 languages, Facebook 91, Google Maps 74, and Google Search 150 (although in 2015 it claimed to support 348). Yet there are more than 7000 living languages in the world. Who's going to decide which ones will make the leap into the digital age?

Any discussion in international venues about 'digital inclusion' or 'digital justice' is meaningless unless this problem is addressed. China understood this 20 years ago and today, along with the USA, it is the only digitally sovereign country on the planet. There is, though, some good news. Recent research by Pimienta (2022) indicates that the languages with most Web content are English with 19.6% and Chinese (21.6%), followed by Spanish (7.9%), and then Arabic, Hindi, Russian, French, and Portuguese (all 3–4%). This suggests that content online has become increasingly multilingual, with the prevalence of English progressively declining from 80% back in 1998 to 20% today. This is great news: the *real Web* is becoming multilingual and multicultural, regardless of all its hegemonic and mainstream *representations*.

3 A UN system co-opted by the interests of global capital

Tim Unwin

Beginning of the preamble to the United Nations Charter

WE THE PEOPLES OF THE UNITED NATIONS DETERMINED

- to save succeeding generations from the scourge of war, which twice in our lifetime has brought untold sorrow to mankind, and
- to reaffirm faith in fundamental human rights, in the dignity and worth of the human person, in the equal rights of men and women and of nations large and small, and
- to establish conditions under which justice and respect for the obligations arising from treaties and other sources of international law can be maintained, and
- to promote social progress and better standards of life in larger freedom. (Signed 26 June 1945, amended 1963, 1965, and 1973)

(UN, n.d.)

The UN Charter originally agreed in 1945 emphasises that the first of its four main concerns was to end the scourge of war. Yet, at the end of 2024 when this chapter was first drafted, horrendous and increasingly drone-driven war was continuing following Russia's invasion of Ukraine in February 2022, Israel was constantly bombarding innocent people living in Gaza and Lebanon following the Hamas attacks on its territory in October 2023, Sudan was torn apart following the resurgence of civil war, the Assad regime in Syria had just been overthrown, and tensions were persisting in the South China Sea between China and neighbouring Pacific countries. By the time the final text was collated, Israel and the USA had also bombed Iran, and the people of Palestine were suffering further massacres as a result of continuing oppression and violence from Israel. We are living through one of the most insecure and volatile periods of recent history, and digital technologies are playing an ever-increasing role in fuelling the scale and extent of these wars, with drones now commonplace, and AI and robotics being combined to create armed assault 'robo wolves' (Parry, 2024). Moreover, the potential existential threat of Artificial Superintelligence (ASI) goes far beyond this, threatening the annihilation of the entire human race (MIRI, n.d.; Yudkowsky and Soares, 2025).

DOI: 10.4324/9781003598008-3

There can be no doubt that the United Nations has largely failed to deliver on its prime role, to save people from the scourge of war. While globally deaths in war declined following the end of the 1939–1945 war and the creation of the UN, they have been increasing significantly since the start of the new millennium, and in 2024 the overall number of conflict deaths, including civilians, rose to a 30-year high (Rustad, 2024). The world is a much less stable and unsafe place than it was in the late 1990s when many believed that the new millennium, with the launch of the Millennium Development Goals (MDGs), would indeed herald a period of peace and prosperity for all. That hope now seems very distant. Many of the global agreements reached and coalitions created in the second half of the twentieth century are now in disarray, and the advent of the second Trump Presidency in the USA in January 2025 has served further to exacerbate this instability. The rules-based international system 'founded on relationships between states and through international institutions and frameworks, with shared rules and agreements on behaviours' (UK, 2018) now seems increasingly fragile.

If the UN did not exist, it would be essential to create something that could indeed fulfil the role its founders envisaged. However, that entity would look very different from the fragmented, competitive, self-important, over-bloated, and increasingly irrelevant body that exists today. The UN system itself is one of the main reasons why digital tech is increasingly being used to drive inequality rather than serving the needs of the world's poorest. Rather than delivering on the four principal reasons for the existence of the UN cited in the epigraph, digital tech is being used to exacerbate warfare, to demean human rights, to fragment justice and respect, to increase social inequalities, and to reduce human freedom. The present global geopolitical crisis nevertheless provides a valuable opportunity for those in the UN system to rethink their roles, and this chapter has been written in part to highlight where some of the most critical digitally-related challenges lie, and propose what needs to be done so that the world's poorest and most marginalised may indeed benefit fairly and appropriately from the use of digital tech.

Why the SDGs and the human rights agenda are not fit for purpose

On the SDGs and digital technologies

Everything looked so different in 2000, and even in 2015 when the Sustainable Development Goals (SDGs) were launched with a fanfare of enthusiasm. The twenty-first century had been born with such hope, despite the scares around the Y2K problem of potential computer errors relating to the formatting of dates (Manion and Evan, 2000). Perhaps we should have seen that as more of a warning of digital challenges to come than we did at the time. Following the claimed successes of the MDGs (McArthur and Rasmusen, 2017), the community of interests represented by the UN system agreed on a massively increased and more complex set of 17 SDGs forming the UN's *Agenda 2023* (UNDESA, 2015) that would run for 15 years from 2015 built around the five 'Ps' of People, Planet, Prosperity, Peace, and Partnership. Many were surprised that this contained no goals around digital tech (or ICTs as they were then known); these were only mentioned in four of the 169 targets (CTO, 2014).

The SDGs themselves remain deeply problematic. The large number of goals (17), targets (169), and indicators (230 unique) has meant that efforts to deliver them have been diffuse and unfocused, with many poorer countries struggling to address them (UNECE, 2020). The process has remained driven by an agenda mainly built upon absolute, rather than relative definitions of poverty, and SDG10 on reducing inequalities has tended to be in tension with many of the other goals that seek to reduce poverty through economic growth. Target setting and measurement have been hugely complex, with the UN itself noting that only 138 of the targets can actually be assessed based on global data and custodian agency analysis (UNSD, 2023). There is a fundamental and unresolved tension between the notion of sustainability (maintaining certain things or processes) and development (about change and progress). The SDGs mainly represent the interests of UN entities and the private sector, rather than the world's poorest and most marginalised people. They have led to an even more bloated 'development industry', not least with the need to try to monitor and evaluate their efficacy (see vignette by Hollow, p. 126). Moreover, despite the large number of the SDGs, they remain focused only on a partial understanding of development that until recently has almost completely ignored digital tech (Unwin, 2015b; 2018).

This failure sufficiently to integrate digital tech within the SDGs led to cross-UN entities such as the World Summit on the Information Society (WSIS) Process rapidly scrabbling in 2015 to put together structures that showed how ICTs could indeed contribute across all the 18 WSIS Action Lines that had been agreed during the original Summit Meetings held in Geneva in 2003 and Tunis in 2005 (ITU, 2015). Interestingly, this showed that only five action lines were relevant to SDG10 on reducing inequality.

Even by the UN's own reckoning, the SDGs were way off target at their mid-point:

> The Sustainable Development Goals Report 2024 makes for sobering reading. It found that only 17 per cent of the SDG targets were on track, nearly half were showing minimal or moderate progress, and progress on over one third had stalled or even regressed.
>
> (UN, 2024c)

As some predicted a decade previously when they were first launched, this should come as no surprise. The failure was inherent when they were conceived (Sneyd, 2024). Only 41 countries, for example, even registered national commitments to SDG transformation by the deadline of September 2023 in response to the UN Secretary-General's plea for contributions to a Rescue Plan for People and Planet (UN, 2023b). For many, though, the response remains that the SDGs are the only game in town and must continue to be supported because there is no alternative (see further on the Summit of the Future). A more sensible strategy would be for the global community to be planning now for the future on what might follow the failed SDGs in 2030, so as to ensure that these discussions do not make the same mistakes as were embedded in the SDGs, and do deliver on the aspirations of the original UN Charter.

The need to balance rights with responsibilities

The Universal Declaration of Human Rights (UDHR) first drafted in 1948 (OHCHR, 1948) is central to the UN's mission and is one of the main underpinnings of the SDGs. The UDHR has also served to underlie much of the opposition by some civil society organisations to what they see as the increasingly problematic development of the digital tech sector over the last two decades. The Association for Progressive Communications (APC), for example, proudly asserts: 'We work for people, especially those facing discrimination and oppression, to have greater power and autonomy through digital technologies to exercise their full range of human rights online and offline' (APC, 2024a). However, although the original formulation of the declaration was driven in large part by a desire to prevent the very real horrors of the 1939–1945 war from ever occurring again, its formulation and evolution remain problematic. Claiming that someone has rights does not mean that they will necessarily ever receive them (Unwin, 2014). Thus, stating that everyone has a right to education (Article 26) says nothing about who will provide the relevant learning experiences. It is therefore essential to balance a 'rights-based agenda' with an explicit 'responsibilities agenda', whereby government agencies, companies, civil society organisations, and individuals are all clear about the structural responsibilities that they should deliver for others (see Chapter 7). O'Neill (2016) refers to these as 'obligations' clearly specified in international law.

I have argued previously that the human rights agenda is deeply problematic (Unwin, 2014; 2017b) for five main reasons (see also Dolinger, 2016; Patrick, 2023): the claim that the rights contained in the document are *universal* is in conflict with a world where moral systems have long varied between cultures in different places; both the UDHR and the ideas behind it were a means through which the interests of a *dominant ideology* have been imposed on others; many states *break their commitment* to the declaration; it is largely based on the rights of *individuals*, rather than the needs of communities; and the human rights agenda has become devalued and diluted, especially so in the context of digital tech (Unwin, 2014).

It is particularly important to stress the interests that underlay and continue to dominate the declaration and its implementation. These were above all the interests of 'modern' global capital that required individual people to be 'free' both to sell their labour power and to have a choice in what they purchase. To this end, it was remarkable that the enlarged drafting committee included representatives from the then 'Communist' states of the Soviet Union and China, but voices of Indigenous peoples, and those from Africa and South Asia, as well as South America (with the exception of Chile) were entirely absent in its formulation. It is often argued that the ethics of reciprocity in the 'golden rule' to do to others as you would have them do to you is the nearest approximation to a universal across all belief systems throughout the world, and that it underlies the declaration's commitment to just treatment, reciprocal responsibility, and justice. However, the golden rule itself remains problematic because none of us knows definitely whether what we would like someone to do to us is actually the same as what they would like us to do to them (see also Chapter 7). It is at best an approximation, and, when elevated to the status of a declaration on rights, can readily serve as an instrument of domination and control. In the name of freedom, it can destroy freedom.

The increasing prominence of digital technologies since the original signing of the UDHR has led to arguments that access to the Internet should itself be considered to be a human right. This is deeply problematic and reflects once more the power of the global digital tech sector in shaping international policy. One of the earliest arguments for such a view can be found in Special Rapporteur La Rue's report to the UN's Human Rights Council in 2011 that expressed concern 'that without Internet access, which facilitates economic development and the enjoyment of a range of human rights, marginalized groups and developing States remain trapped in a disadvantaged situation, thereby perpetuating inequality both within and between States' (UNGA, 2011, p. 17). In particular, since the human right of education is increasingly being delivered through the Internet, it is often argued that access to the Internet itself should become a human right, a view that is advocated particularly vehemently by the so-called EdTech sector. If access to digital technology is accepted as a human right, then EdTech companies like to insist that others, especially governments, should fund it so they can then use the Internet to extract profit. However, this is nothing like the rights to life, liberty, and security of person (Article 3), or to food, clothing, and housing (Article 25) that were at the heart of the original UDHR. Moreover, there are strong grounds for arguing that the 'right to be unconnected' is even more important than any putative right to be connected. Being unconnected, we remain free from the powerful clutches of the Digital Barons.

Nevertheless, and more positively, the growth of interest in AI since the mid-2010s has highlighted increasingly prominently both the positives and negatives that the use of digital tech poses for human rights. As the OECD (2023) has neatly summarised:

> The digital environment offers tremendous opportunities to enhance people's ability to exercise their human rights, but it can also create new and exacerbated risks, as well as new links and tensions between rights. For example, efforts to safeguard rights associated with online safety by removing harmful content online like hate speech may be seen as interfering with others' freedom of expression.

The ongoing debate around the positives and negatives of digital tech use for human rights can be seen as a welcome development, and one of the grounds for optimism expressed by proponents of the Global Digital Compact (GDC) is that it did indeed provide a platform where civil society was able to advocate for the inclusion of more dimensions of human rights than has heretofore been the case (Finlay and Betancourt, 2024). Nevertheless, it is essential to develop a much more robust system of balances focused around responsibilities as well as rights, which rigorously identifies exactly what the specific responsibilities are for public sector, private sector, and civil society entities if the human rights agenda is to be delivered. Chapter 7 offers one possible framework for such an agenda.

The mirage of multistakeholder partnerships

The notion of multistakeholderism within the international system has also tended to serve the interests of the private sector alongside the governments of richer and more powerful countries, more than it has the world's poorest and most marginalised (Unwin, 2017a). Such partnerships are almost unanimously seen as being a 'good thing', and essential for any global discussions around digital tech and development. This section challenges this assumption, and argues instead that we need to shift the balance away from loosely defined multistakeholderism to a clearer and tighter framework built on truly multi-*sector* partnerships that explicitly bring together the interests of the public sector, the private sector, and civil society equally.

Partnerships and the subversion of the UN by private capital

Multistakeholderism in the international system emerged from the wider neoliberal agenda of Public-Private Partnerships (PPPs) that began to be implemented in the 1980s and became increasingly popular during the 1990s as governments, especially in Europe and North America, sought to encourage private capital investment into erstwhile public activities, while also reducing the political risk; governments could always blame the private sector if things went wrong (Unwin, 2015a). Although these have subsequently been widely criticised (Blavot *et al.*, 2022), enthusiasm for them at the time that the ICT sector was also beginning to flex its muscles was remarkably high. Most UN agencies had, and increasingly still have, insufficient funds to fulfil the growing ambitions of their member states and their leaders in the first two decades of the twenty-first century, with total UN funding only rising from US$41.5 billion in 2011 to US$68.5 billion in 2023 (CEB, 2024). This created considerable opportunities for an expansionary private sector to deliver new initiatives in partnership with governments and UN agencies. For the companies involved, this provided an ideal opportunity to influence global policy while also getting to know and influence national politicians and civil servants who could be 'persuaded' into purchasing their products (see Note Three in Chapter 1). Although the private sector only officially contributed 4% of the UN system's funding in 2019 (Topping *et al.*, 2021), its real contribution to initiatives on the ground and sponsorship of UN events is many times larger. By the end of 2023, the private sector, for example, had committed two-thirds of the total pledges to the Partner2Connect initiative valued at US$36.78 billion (ITU, 2023).

The dangers of this process were already highlighted in a prescient paper two decades ago by Martens (2007), who summarised many of the key risks of such partnerships for the UN. He also emphasised that global corporations were already more wealthy and powerful in shaping international policy than were most sovereign states, especially those in the poorest parts of the world (see also the vignette by Fiormonte, p. 40). By 2021, the market capitalisation of Apple alone was US$2.1 trillion, which was more than the GDP of all but the seven richest countries in the world (Wallach, 2021). Such figures emphasise the scale of difference between the economic, and thus also political, power of the major global corporations and the relative weakness of most sovereign states. This is especially so with respect to the major digital tech companies that have been essential for driving economic growth and have always played an important role in international organisations such as the International Telecommunication Union (ITU) and the Commonwealth Telecommunications Organisation (CTO). There is nothing exceptional about this, but they can never deliver alone on the needs of the world's poorest. Governments, civil society, and individuals should never expect them to. These companies drive inequalities, and if this is to change, they ultimately need to be held responsible to governments that theoretically hold the essential power of redistributive justice in the interests of all their citizens. The potential of such partnerships in the digital tech sector was well realised in the original UN World Summit on the Information Society (WSIS) held in Geneva in 2003 and in Tunis in 2005, which were open not only to governments and UN bodies, but also to other international organisations, NGOs, and civil society, and the business sector. Civil society representatives actually outnumbered those from the business (private) sector (ITU, 2006), but there was a feeling among many of them that the private sector was already too heavily involved in shaping the WSIS agendas and conclusions, and that the voices of civil society were insufficiently heard. The last 20 years have shown just how much the private sector has since come to dominate global digital interventions and policy making. Despite this, many of the Association for Progressive Communications' (APC) proposals at the time of the original WSIS summits (APC, 2005) have indeed come to pass, especially with the creation of the Internet Governance Forum (IGF) which first met in Athens in 2006 as a discussion forum bringing together all stakeholders with an interest in Internet governance, but without any formal or binding decision-making process (Andjelkovic, 2007; digwatch, 2024).

There are also many challenges with the implementation of so-called multistakeholder partnerships on the ground. All too often, the word 'partnership' is used merely to describe a loose relationship between several organisations implementing a project or intervention, without there being any formal structure to ensure its success. It is hardly surprising therefore that these often fail to deliver on their rhetoric (Lallana's vignette on p. 49 provides a fascinating insight from education initiatives in the Philippines). Well-established and excellent models of good practices in digital tech and development partnership are all too often ignored (see Tennyson, 2003; Unwin 2005b; 2015a; Unwin and Wong, 2012; Thorpe *et al.*, 2021). Why is it that such frameworks are so rarely used, and those creating new initiatives fail to draw on past experience and wisdom? Some of the reasons for this are explored further in Chapter 4, but three are worthy of note here: (1) often those involved are in too much of a hurry, and the word 'partner' seems a simple way to capture their relationship; (2) development partnerships are in vogue, and crediting the relationship with such nomenclature elevates its status; and (3) those involved simply do not do their homework properly. As a result, many such digitally-focused partnerships disappoint in terms of delivering real development outcomes in the interests of the poor. Good guidance on partnerships nevertheless exists, and often focuses on: the need for a formal document agreeing aims, objectives, intended benefits, and contributions of all partners; having a capable and well-staffed management office; engaging as many partners as possible at the beginning to shape the initiative, rather than accruing further partners later; grounding it in a transparent and open ethical framework; maintaining a clear and realistic resourcing structure; constant monitoring, evaluation and reporting to all partners; and having an influential high level champion (Unwin, 2015a; see also vignette by Hollow, p. 126). Too many 'partnership' initiatives developed by UN agencies to implement digital tech for development interventions do not have all of these in place, and consequently under-deliver.

Reinventing travail: ICT in basic education in the Philippines

Emmanuel C. Lallana

(Chief Executive, ideacorp and Professorial Fellow, University of the Philippines)

The use of ICT in education in the Philippines began in the mid-1990s, supported by many private sector and civil society organisations keen, as elsewhere in the world, to see digital tech as a means of improving teaching and learning, education management, and preparing learners for the future. However, it has been beset by problems that continue to hound the programme. At the heart of these has been the inability of the Department of Education (DepED) to deliver on its Computerisation Programme (DCP), launched in 2010. A 2025 cut in the computerisation budget by P10 billion (€154 million) was justified by a legislator who said that 'Congress cannot keep throwing good money after bad'.

This results in part from poor programme management, irregularities in procurement, and equipment becoming obsolete. Another well-known cause of the lack of sustained forward movement in ICT4D is the habit of hitting the reset button when a new President and/or Secretary of Education takes office. However, there are also deeper reasons, most importantly that the initial programmes were designed to train students in ICT literacy to make them fit to 'meet the challenges of the 21st century'. While relatively successful in its ICT literacy efforts, DepEd faced difficulty in integrating digital tech throughout teaching and learning. Teachers especially need to be well trained in the use of digital tech, and yet only two of the 37 strands of the Professional Standards for Teachers (PPST) refer specifically to ICT. There have, though, been some successes, with the eSkwela programme launched in the early 2000s, and using digital tech to support learning by out-of-school youth and adults (Unwin *et al.*, 2017). This won several international awards but quietly died when project leadership was turned over to DepEd, although some eSkwela modules are reportedly still being used in some regions of the country.

Government has to better understand the shifting education needs, ensure alignment in its technology-learning framework, and improve its strategic planning and adaptation, if it is to improve its educational technology decisions. Otherwise, the significant benefits of harnessing ICT for education will never be fully realised.

Replacing multistakeholderism with overtly multi-sector structures

Strictly speaking, the notion of 'multistakeholderism' merely means involving many different stakeholders. In reality, different organisations espouse different structures through which this can be achieved. The IGF, the WSIS process, and ICANN (the Internet Corporation for Assigned Names and Numbers) all proclaim multistakeholderism, yet they do so in subtly different ways that reflect their varying purposes and interests. Multistakeholderism is a sufficiently fluid concept that most can sign up to without really thinking about it too hard or worrying too much.

However, not only has the private sector come to dominate many digital multistakeholder processes and interventions through its financial power and influence, but it has also achieved this more subtly through its indirect influence over many civil society entities and indeed governments. The very essence of global multistakeholder events and activities also tends to be exclusionary. The poorest and the most marginalised rarely attend the numerous digital multistakeholder events held every year across the world on similar themes because they cannot afford or otherwise have the capacity to do so. Moreover, the civil society organisations that claim to represent them also have other agendas and interests in participating (see also Chapter 4; Unwin, 2017a). Governments of small and poor states neither have the number of officials nor the wealth to attend all the relevant global events, and often ponder about which ones they should attend, if at all. People with disabilities are also vastly under-represented in multistakeholder meetings relating to digital tech. Certain topics are also excluded. For example, the word 'gender' within the UN system is all too often used as a synonym for 'women' or differences between men as in the 'gender digital divide' (Buskens and Webb, 2014; see vignettes by Okello, p. 51, and Sterling, p. 66); issues around the use and abuse of digital tech within LGBTQ/LGBTIQ/LGBTQI communities are all too frequently ignored, in part because of the strongly different policies of states within the UN relating to sexual orientation. To be sure, many international organisations have sought to grapple with some of these issues, especially through the introduction of online participation during and since the COVID-19 pandemic, but it still remains difficult to interact effectively through existing online conferencing systems and these require stable, high-quality Internet connectivity to which poor and marginalised individuals and communities frequently do not have access. They also cannot yet replace the crucial personal networking that happens on the ground during such events.

The notion that socio-economic structures can broadly be divided into three sectors, namely the public sector (government), the private sector (business), and the third sector (civil society) originated in antiquity, but came to the fore especially in

European political thought during the seventeenth and eighteenth centuries (Kenny *et al.*, 2016). Each sector in an effective and functional society has specific roles and responsibilities, with reciprocity between them: governments should legislate and regulate in the interests of the people; businesses produce goods and services for profit; and civil society serves independently of both to advance the interests and will of citizens, especially by holding governments to account (see further in Chapter 7). This framing has come to dominate global discourse relating to multi-stakeholderism, but it is a distinctly European notion, and it would be interesting to explore alternative traditional cultural constructions of collaboration from Africa, Asia, and South America to a greater extent to see what they might also have to offer. The key point about this for the present chapter is that having 'many stake-holders' (multistakeholderism) does not necessarily mean that representatives of all three of the sectors must be present in a balanced way. Hence, the term can be, and indeed often is, used when many stakeholders largely from governments and the private sector alone are involved. For the success of such partnerships and discussions, though, it is essential also to involve civil society as a partner of equal worth. Civil society has a fundamentally important role in keeping the excesses of both governments and businesses in check, and during the neoliberal epoch has also been of great importance in delivering interventions on the ground to benefit those on the margins of society that many governments no longer see as their responsibility (see Okello's vignette, p. 51). Many staff and volunteers in civil society organisations have also gained considerable knowledge from their experiences in working with the poor and marginalised on the ground, which is crucial for a more holistic understanding of how digital tech interventions can best be implemented in practice in the interests of all, rather than the few.

As the next section illustrates, systems established globally to develop policies at the interface between digital tech and development under the rubric of multistakeholderism may consult civil society, but all too often they are led and shaped primarily by the private sector and governments. Multistakeholderism is often a mirage deliberately constructed to disguise the perpetuation of what in effect remain Public-Private Partnerships. It is essential therefore to shift our language to the term 'multi-sector', explicitly involving civil society as an equal partner, if we hope to have any chance of benefitting the poorest and those at the peripheries of our existing societies through the design and use of digital tech.

Is it really more of the same since WOUGNET's inception in 2000?

Dorothy Okello

(Associate Professor and Dean, School of Engineering, Makerere University, Uganda)

As Women of Uganda Network (WOUGNET, n.d) marks her 25th birthday, it is opportune to step back and reflect. With a mission to promote and support the use of information and communication technologies

(ICTs) among women and women's organisations, WOUGNET has grown into a critical actor promoting gender inclusive digital development in Uganda. Where 25 years ago, access and use of ICTs were the core issue, WOUGNET's mandate has expanded to address digital rights, civic tech, policy advocacy, and online safety.

Why then does it feel like we are still talking about the very same kind of issues? Overall, Internet access has risen to 43% in 2024 (Uganda Communications Commission, 2024) from a paltry 1% in 2000 (ITU, 2000), but women in Uganda are 47% less likely than men to use the mobile Internet (GSMA 2024c). Furthermore, 42% of women report being unaware of the mobile Internet, compared to 27% of men. This is despite the mobile ownership gap having closed to about 11%, and 20% of women owning smartphones compared to 29% of men. Drivers for these gaps, such as affordability, literacy and digital skills, and safety and security, remain key barriers for women – and for men too (GSMA, 2024d).

While WOUGNET has responded by implementing digital literacy campaigns, policy dialogues, and advocacy around online gender-based violence, digital inequalities continue to widen, particularly, for women and young girls in rural communities and/or with disabilities. Already with socio-economic disadvantages, they stand the risk of becoming digitally invisible in a fast-evolving digital economy – driven by our Digital Uganda Vision that seeks to empower its citizens and to strive towards the goals of universal inclusion, sustainable development, economic progress, and poverty eradication through digital innovation across multiple sectors (Nabbale *et al.*, 2023).

Even as the digital system has matured over the years, so too have the inequalities within it. WOUGNET's journey reveals that while technology holds transformative potential, there is still need for targeted policies, inclusive infrastructure, and gender-responsive digital education so as not to reinforce – and even widen – existing social divides. The journey of a thousand miles does start with one step, and we all have a role to play.

Power struggles within the UN system and the emergence of the Global Digital Compact

The UN system is complex, consisting of six main bodies (the Secretariat, the General Assembly, the Security Council, the Economic and Social Council, the Trusteeship Council, and the International Court of Justice), six Funds and Programmes (including UNDP and UNICEF), its 15 specialised agencies which function as autonomous international organisations (including the ILO, the ITU, UNESCO and the World Bank), nine other entities and bodies (such as UNHCR, the UN University, and UN Women), and eight related organisations (such as the IAEA and the WTO) (UN, 2024d). Relationships between them all vary, depending on their remits, the amount of funding from their members and partners, and the personalities of their leaders and senior management teams (see also Chapter 5 on the 'me' syndrome).

Digital tech companies permeate the UN system

From the late 1990s, small numbers of staff in many UN agencies and the central Secretariat started taking greater interest in how digital tech might be used to enhance communication within the UN system and for the delivery of its information and programmes. The two WSIS Summits in 2003 and 2005 and the rapid spread of mobile telephony, especially following the launch of 3G in the mid-2000s, added further impetus for the widespread integration of digital tech within the UN system. However, there was relatively little real expertise in digital tech within most of the UN at that time, and the rapid pace of innovation provided an incentive for agencies increasingly to turn to the private sector for advice and assistance, as well as for additional funding at a time when resources from member states were under greater pressure as agencies sought to deliver their expanded contributions to the MDGs.

Within the telecommunications space the ITU was unusual for a UN agency in already involving many private sector members from its historical roots, and also in having the greatest technical expertise in the field of ICTs. Hence, much of the work of their Study Groups was already heavily influenced by the views of the private sector. The WSIS process had led to eight UN entities (the FAO, the ILO, the ITU, UNCTAD, UN DESA, UNESCO, the WHO, and the WMO), each taking a lead in carrying forward one or more of the Action Lines and reporting to the annual WSIS Forums in Geneva. Gradually, as they developed their wider activities, two things happened. First, the boundaries between the activities of different agencies began to blur, bringing them into potentially greater competition for resources, private sector engagement, and areas of mandate. Second, digital tech companies began to take a much more active role in participating not only in the global UN events, such as WSIS and the IGF, but also in other technology-related forums increasingly being organised by specific agencies, such as UNESCO's Mobile Learning Week. The formulation of the SDGs, launched in 2015 with their home in UN DESA, provided further impetus for digital tech companies to emphasise their relevance to global sustainable development agendas by working together with UN agencies to help deliver their objectives. Such participation, often through financial or in-kind sponsorship, provided them with good networking opportunities, a chance to meet high-level government officials and influence national policies, as well as an opportunity to expand their markets by engaging in interventions on the ground under UN auspices. Above all, it was good for their corporate image. For UN agencies, such informal partnerships helped increase the funding available for the interventions that their leaders and members wished to implement, and also ensured that they could benefit from using some of the latest technological innovations.

This increased engagement of the private sector in international development initiatives at a global and national scale was also taking place much more widely through direct business relationships between governments and the private sector, and also through the mediation of foundations and other international forums. The World Economic Forum thus convened a novel Global Education Initiative (GEI) between 2003 and 2011, which involved companies, governments, civil society organisations, and academics, to implement educational activities through digital tech in Jordan, Egypt, and Rajasthan, and was at the forefront of implementing successful digital partnerships at that time (Unwin and Wong, 2012). Its Steering Board consisted of 20 of the most important digital tech companies, including Cisco, Intel, and Microsoft.

Meanwhile, in parallel with the annual WSIS Forum events, the IGF meetings convened by the UN Secretary-General, guided by his appointees on the Multilateral Advisory Group (MAG), were also being held annually in countries able and willing to fund them. The MAG seeks to have approximately equivalent regional representation, with around 40% government members and then an equal split between the private sector, civil society, and the technical community. Frequently, though, members of the technical community as well as government representatives are also people who are, or have previously worked in, the private sector. The IGF also established a series of national and regional forums to discuss matters of Internet governance and actively involved youth (see vignettes by Morenets, p. 55 and Longmore, p. 27). However, it was set up primarily as a discussion forum, and was never intended to drive forward action and progress on the issues discussed. Moreover, the same people have often attended and contributed to all three IGF, ICANN, and WSIS annual meetings, and there was frequently considerable overlap of discussions and content between these forums.

There have also been a plethora of other international multistakeholder and specialist academic conferences on ICTs and Development beyond the UN system, not least those convened by the Institute of Electrical and Electronics Engineers (IEEE), the International Conferences on Information & Communication Technologies and Development (ICTD), first held in 2006, and the ICT4D multistakeholder conferences convened by Catholic Relief Services (CRS) since 2010, most of which had significant private sector involvement as sponsors and speakers, especially from the USA, and more recently China. To these can be added the many digital and development conferences arranged by regional organisations, such as the Caribbean Telecommunications Union, or the series of Asia-Pacific Conferences on Communications since 1993, and the Netmundial events in 2014 and 2024 convened by Brazil, which claimed to be the first multistakeholder gathering of its kind contributing to the evolution of Internet governance. Alongside these, there are numerous private sector exhibitions and events, such as the GSMA's Mobile World Congress held annually in Barcelona since 2006 with its Ministerial Programme claiming to be the 'world's leading forum for policymakers and global industry leaders from across the mobile ecosystem and adjacent industries' (GSMA, 2024b). The overlap, duplication of effort, and considerable expense of running such events, during which it is often difficult for the world's poorest and most marginalised to have a voice, even if they are able to attend, seems primarily to serve the interests of the richer and more powerful companies and governments. To this end, it is also important to note that since the middle of the 2010s, the donor landscape relating to digital tech has changed significantly, and new governments, especially from hydrocarbon-rich states, have increasingly sought to raise their profiles and influence within the global system by responding to the need by UN agencies for funding, through hosting or sponsoring their major international gatherings (see Chapter 4).

Beyond the token seat: Rethinking youth inclusion in global forums

Yuliya Morenets

(Barrister, digital governance, Youth IGF founder, cybersecurity advisor, inclusion champion)

When visibility is mistaken for empowerment, youth participation risks becoming performance; a reflection drawn from one real exchange.

A few years ago, I was waiting for a friend in the registration lobby during an Internet Governance Forum meeting, when a conversation nearby caught my interest. It was between three young participants, and revealed much about the realities of youth engagement in global spaces:

Ali: I put my name on the speaker list, but no one contacted me. I came here on my own – took a small loan to afford the ticket. I just wanted to share my ideas.

Nanna: You paid to come? We're usually funded. It's a great opportunity – many of us found jobs this way.

Ali: But shouldn't it be about our communities, not just careers?

Nanna: Honestly, it's about visibility. You have to be known.

Ali: Isn't that the system we're meant to change?

Malik: I got in through my national Youth IGF. I stayed involved, people noticed. That's how I got invited. It's not always about what you know, but who knows you.

Ali: So we repeat the same cycle – recognising the known instead of making space for the new?

Nanna: The issue is we lack a pipeline. Young people are invited to speak, but not supported to act. Their ideas are raw, but often unshaped.

This brief exchange reflects a deeper truth: youth inclusion often rewards familiarity over innovation. The same names are recycled, while others are left out – not for lack of ideas, but because they don't fit the mould. If we want real participation, we must move beyond tokenism. We need to invest in youth not just as performers, but as builders – messy, passionate, and still learning. The point is not to polish every voice but to hear the ones that haven't been heard.

Let's stop asking who fits. Let's ask who's ready to try – even if they haven't spoken before.

The role of the UN Secretariat: towards the Global Digital Compact

By the late 2010s, the pace of technological innovation, combined with the increasing duplication and competition between UN agencies for resources and control over the 'digital and development' space, were exacerbating the already complex and cumbersome UN system that was seeking to reach global agreements on digital tech and to deliver effective development interventions. This was an opportunity for ambitious individuals and the major tech companies to gain an even greater hold over the UN system's approach to the use of digital tech by going straight to the central UN Secretariat and the new Secretary-General Guterres (appointed in 2016), who had earlier in his life studied physics and electrical engineering at the Instituto Superior Técnico in Lisbon, and had served previously for a decade as UN High Commissioner for Refugees.

One of his early acts was to create a High-Level Panel on Digital Cooperation (HLPDC) in 2018 to consider 'the ways we work together to address the social, ethical, legal and economic impact of digital technologies in order to maximise their benefits and minimise their harm' (UNSG HLP, 2019, p. 4; see also UN 2020). Instead of supporting existing systems and agencies within and beyond the UN, he was persuaded to embark on an entirely new approach that largely duplicated work that had already been done over the previous decade by other entities. Very significantly, the HLPDC, chaired by Melinda Gates (then Co-Chair of the Bill and Melinda Gates Foundation) and Jack Ma (Executive Chair of the Alibaba Group), was heavily dominated by members from companies, and scientists with close links to the private sector (similar also to the composition of the Secretary-General's AI Advisory Board, established in 2023, and discussed in Chapter 1). Of the 20 members of the panel, only three had any real experience of international development on the ground. Very few even had practical experience of working with and serving the interests of the poor. The panel sought to consult widely, but such conversations largely repeated what had been said over the last decade in many other existing forums (see vignette by Okello, p. 51). There was little new in its findings. Nevertheless, it was positive to note that the panel's first report in 2019 did mention inequality/inequalities eight times in the text (UNSG HLP, 2019), although the word equity was noticeably absent.

The Secretary-General's subsequent *Roadmap for Digital Cooperation* (UN, 2020), based in part on further widespread consultation, did indeed begin to address some of the harms caused by the use of digital tech, but it still only mentioned inequality/inequalities seven times. It also failed satisfactorily to address the root causes of the harms caused by the use of digital tech, primarily because it was based largely on a belief in digital instrumentalism and determinism. Its focus was mainly on the role digital tech should play in driving economic growth for the benefit of humanity, rather than explicitly addressing ways that the harms caused by its use can be mitigated. This is well captured in Guterres' (2020b) own statement that 'Digital technology is shaping history'. Such assertions fail sufficiently to recognise that it is the people who conceive, design, and sell the digital technologies who are shaping history, and who therefore need to be encouraged to change their ways if digital tech is truly to be used to serve the interests of humanity more equitably.

Building on this *Roadmap*, Guterres then swiftly chose to appoint his first Tech Envoy, Fabrizio Hochschild, despite there being unresolved alleged complaints about his previous time as a UN Under Secretary-General and Special Advisor

(Heikkilä, 2021; UN, 2021b; BBC, 2022; Kotecha, 2022). Significantly, Guterres chose someone for this appointment who did not have experience in the broad field of digital tech, but was rather a senior UN civil servant who had served as his Special Advisor on the state of the world and 'our common future' from 2019–2021, having previously led system-wide strategic processes in his Executive Office from 2017–2019. Prior to this, Hochschild had mainly gained experience in working with UN agencies involved in human rights and refugees. As the new Tech Envoy clearly acknowledged in one of his first tweets, he was 'a relative newcomer to the field' (Hochschild, 2021; Unwin, 2021), and it seems clear that Guterres had chosen to appoint him for his organisational experiences rather than for his knowledge of digital tech and development. Shortly after his appointment, however, Hochschild was placed on paid leave pending an investigation, and was subsequently relieved of his duties despite concerns expressed by some of his colleagues that 'the internal investigation process, which we trusted would be fair, has been heavily influenced by other factors, including a parallel media campaign filled with untrue allegations against Fabrizio' (Aibinu *et al*., 2022). This sorry incident highlights a wider and important characteristic of the UN Secretariat, especially in the arena of digital tech for development, where a lack of relevant expertise and personal connections are often key features of those appointed to leadership or senior advisory positions.

In September 2021, the UN Secretary-General then produced his next report *Our Common Agenda* (UN, 2021), which was a somewhat rambling, aspirational document that served as an attempt to salvage something from the impending wreckage of Agenda 2030 and the SDGs. The seventh of its 12 commitments was on improving digital cooperation, and slimmed down the earlier list of issues in the *Roadmap* (UN, 2020) to seven key proposals forming an agenda for a proposed new Global Digital Compact:

1. Connect all people to the internet, including all schools.
2. Avoid internet fragmentation.
3. Protect data.
4. Apply human rights online.
5. Introduce accountability criteria for discrimination and misleading content.
6. Promote regulation of artificial intelligence.
7. Digital commons as a global public good.

This largely seems to have reflected the interests of global capital and the advocates of neoliberalism, the rich and powerful states, and para-statal entities, and those of the UN and its agencies. On the positive side, though, it did highlight a few of the potential harms caused by using digital tech and raised questions surrounding the rise of large tech companies as geopolitical actors. As *Our Common Agenda* thus noted, 'Currently the potential harms of the digital domain risk overshadowing its benefits' (UN, 2021, p. 62), and 'Serious and urgent ethical, social and regulatory questions confront us, including . . . the emergence of large technology companies as geopolitical actors and arbiters of difficult social questions without the responsibilities commensurate with their outsized profits' (UN, 2021, pp. 62–63). These are undoubtedly welcome assertions, although some 20 years too late. The horse had already bolted.

During this period, direction from the interim leader of the Office of the Tech Envoy and drawing on the proposals in *Our Common Agenda* encouraged greater emphasis on UN entities working more closely together on digital tech and development. Thus, UNDP and the ITU combined their strengths to create a Joint Facility for Global Digital Capacity in 2021 (UNDP, 2022). Likewise, the UNOSET and the ITU began to try to rectify the omission of objectives and targets relating to digital tech in the original SDGs, and in 2022 formulated a set of Aspirational Targets for 2030 that aimed to achieve universal and meaningful digital connectivity in the 'decade of action' in line with delivering on the SDGs (ITU, 2022; UNOSET, 2022). Although well intentioned, these suffer from many of the previous criticisms of such efforts already outlined in this book: they mainly serve the economic interests of the private sector rather than the world's poorest and most marginalised; they are idealistic, and highly unlikely to be achieved; they contain nothing about reducing digital harms; there is nothing in them about accessibility for people with disabilities; and by claiming universality, they appear to abrogate the right to be unconnected. Perhaps not unsurprisingly, these targets did not manage to find their way explicitly into the GDC (although see para. 11a), despite all the effort that had gone into producing them. To reinforce the argument about private sector engagement, almost all the contributors to the UNOSET (2022, p. 21) report on setting targets for achieving universal and meaningful digital connectivity were from UN agencies, private sector companies or industry bodies/foundations, with the only others being a Mexican government representative and a US academic who had previously worked for his government in the telecommunications sector.

Eventually, in July 2022, A. S. Gill, one of the two Executive Directors of the Secretariat of the High-Level Panel on Digital Cooperation (HLPDC) was appointed as the new Tech Envoy. He had formerly served in Indian diplomatic positions, was India's Ambassador to the Conference on Disarmament, and professed an interest in Artificial Intelligence, but was by no means an expert in delivering digital tech interventions on the ground or for poor people. With this background he had the challenge of driving forward the UN Secretary-General's agenda of creating a Global Digital Compact (GDC), building on the work of the HLPDC. This process continued to replicate the discussions already held over the previous two decades since the original WSIS Summits, and, despite its widespread consultations, was deeply flawed: first, it was a reflection of the ways in which the ideologies and practices of digital tech companies had come to dominate UN rhetoric around digital tech; second, the issues that it addressed, the questions it asked, and the ways in which the consultation were constructed, largely served the interests of those companies and powerful states, rather than those of the world's poorest and most marginalised individuals and communities; and, third, it paid scant attention to the future relationships between machines and humans, the environmental harms caused by the design and use of digital tech, and the increasing enslavement of the majority of the world's people through the activities of digital tech companies of all sizes (Unwin, 2023a). In essence, the consultations were framed around the seven

Figure 3.1 Contributions to the online consultations on the Global Digital Compact

Source: https://www.un.org/techenvoy/global-digital-compact/submissions (accessed 5 December 2024).

proposed digital areas and themes in the already flawed *Our Common Agenda*, and although some 400 additional inputs were received, it was this framing, which private sector interests had been so influential in producing, that shaped the GDC. If there is any doubt over where the power lies, this image on the Tech Envoy's site, noting the US$5tn+ market cap of the private sector contributors to the consultation as one of its three highlights, says it all (Figure 3.1).

In 2023, the ITU and the UNDP in collaboration with the Inter-American Development Bank (IDB) and the Boston Consulting Group (BCG) then proposed an ambitious SDG Digital Acceleration Agenda, claiming that 'digital can play three crucial roles: as a core channel, and platform, for the delivery of vital services; as a tool to deliver data and insights to achieve each of the SDGs, and as a catalyst for wider progress' (ITU and UNDP, 2023). In so doing, they also claimed that digital technologies can benefit 70% of SDG targets, and they highlight 34 digital solutions that are helping to advance them. This approach is once again a clear example of the sort of digital determinism and instrumentalism that drives so much of the UN's rhetoric around digital tech, and therefore remains part of the problem and not a solution. It can also be interpreted as a clear attempt by the leadership of the ITU and UNDP to be seen to be collaborating, drawing on the synergies between the ITU's specialist technical capabilities and the UNDP's development experience on the ground, in the face of the UN Secretary-General's approach to creating the Global Digital Compact.

The Summit of the Future and the Global Digital Compact

There have been numerous assessments of the Global Digital Compact that was eventually agreed during the Summit of the Future in 2024, most of which have been broadly positive, with many of its critics even being able to find something to like about it (APC, 2024b; Gaballa, 2024; Teleanu, 2024). For some, it is a good compromise, including much of the previously acceptable rhetoric around the importance of a human rights-based approach, gender equality, multistakeholder governance, and concerns over emerging technologies. However, these have not well served the interests of the world's poorest and most marginalised in the past, and there seems little reason to suggest they will do so in the future, especially given the powerful interests underlying the processes giving rise to the GDC in the first place. This is despite the good intentions of many committed and experienced people working within the UN system.

The Summit of the Future was widely seen from within the UN system as being a success, leading to agreement on the Pact for the Future, the Global Digital Compact, and a Declaration on Future Generations which were adopted in Resolution 79/1 of the UN General Assembly on 22 September 2024. The President and CEO of the United Nations Foundation thus claimed, 'Under pressure, the world has chosen the path of cooperation' and this enabled UN Member States 'to find common ground in the face of conflict, polarization, and distrust' (Cousens, 2024). Yet, remarkably few of the world's people know anything about the Summit, and its agreements sadly continue to reflect the increasing irrelevance of the UN and its processes for achieving positive change in the world. Many of those who do know about it see it as a vanity project conceived by Guterres as he nears the end of his second term of office in 2026. It includes little about fundamental reform of the UN on the scale that is required, or about how the systems that it promotes can really deliver on poverty reduction from a relativist perspective that could lead to real empowerment of the world's poorest and most marginalised people. To be sure, there was a specific focus on youth and future generations (see vignette by Longmore, p. 27) but in practice this is likely to continue to be similar to previous approaches through which youth representation tends to be paternalistic and largely replicates existing political structures (see vignette by Morenets, p. 55). The voices of young people at risk of living and working on the streets are largely absent. Sadly, there was little fundamentally new coming out of the Summit, and, more worryingly, even less on practical and realistic ways through which its aspirations can be achieved; its 56 Actions mostly remain vague, imprecise, and non-binding. The Summit's greatest achievement seems to have been that it shored up a commitment to multilateralism, and did try to create consensus in an increasingly fragmented world. As Rigg (2024) commented, despite its failure 'to deliver a clear vision of the future that we want, or to provide a binding roadmap on how to get there . . . Multilateralism was brought "back from the brink" and the UN was taken off life support.' Was the very considerable overall cost of preparing for and convening the Summit of the Future really worth so little?

The first draft of the GDC itself was deeply problematic, and many of its basic problems survived into the final approved version of the text. Above all, both combined an instrumental and deterministic view of digital tech, and both saw it as providing solutions to the deeply flawed SDG agenda. This is well captured in its opening paragraph discussed in Chapter 1 (UNOSET, 2024). Such overwhelmingly positive rhetoric fails sufficiently to recognise that it is the people designing and selling digital technologies who are driving the transforming in their own interests, and the document continues to promote the view that digital tech can solve the problem of achieving the deeply problematic SDGs that will in practice never sufficiently deliver on the needs of the world's poorest and most marginalised. Moreover, this opening paragraph also makes no mention of the harms caused by digital tech, although these do subsequently feature six times further down in the document. Likewise, the paragraph appears to suggest that digital technologies are good for the planet, and does not sufficiently recognise the immense damage that they are causing to nature, although concerns over the physical environment are indeed noted briefly four times in the subsequent text.

The GDC's five main objectives are unsurprising, given that they are based on an attempt at widespread consultation and are closely in line with much previous rhetoric over the last two decades:

1. Close all digital divides and accelerate progress across the Sustainable Development Goals.
2. Expand inclusion in and benefits from the digital economy for all.
3. Foster an inclusive, open, safe and secure digital space that respects, protects and promote human rights.
4. Advance responsible, equitable and interoperable data governance approaches.
5. Enhance international governance of artificial intelligence for the benefit of humanity.

(UNOSET, 2024)

These beg the question once again as to why so much expense and time had to be spent on drafting and agreeing them. Perhaps this was necessary so that staff at UNOSET and the UN Secretariat more generally could actually learn something about digital tech and development (see also Chapter 2 for a discussion of terminology, such as 'digital divide'); very few of the staff at UNOSET had any real previous experience at the interface between digital tech and development. It is important to stress, though, that despite this, the overall document does nevertheless serve as a positive summary of much accepted knowledge. In particular, it is good to see at least some recognition of the risks of digital tech, the need for greater attention to be paid to online safety, the importance of affordability and accessibility, its underlying foundation in human rights (mentioned 29 times), its mention of gender-based online abuse and harassment, the challenges of data privacy, and the need for systems of governance relating to emerging technologies (Unwin, 2024b). The problem is that this existing acceptable knowledge has not yet reduced the inequalities caused by those who design and sell digital tech.

The GDC is remarkably weak on precise responsibilities and on practical actions for delivery. In contrast to the frequent discussion of human rights, the words 'responsible/responsibility/responsibilities' are thus only mentioned 15 times, and then only vaguely in terms of undefined respective roles and responsibilities of different stakeholders; 'obligations' are only mentioned twice. Signatory governments have likewise agreed to eight 'must', 42 'will', and 24 'committed/commitment' statements, but many of these are not time-bound, and those that do mention a date simply reflect aspirations to deliver by 2030. Most of them have no clear indication as to how they might be delivered and, as with the SDGs, their sheer number will make it difficult, if not impossible, for governments of the poorest and smallest countries to comply. The vagueness of the document is typified by paragraph 67, which states, 'We invite private sector and philanthropic actors to consider financial pledges in support of the implementation of this Compact' (UN 2024b, p. 14). Nevertheless, optimists such as Finlay and Betancourt (2024) suggest that this vagueness may actually be a positive opportunity for civil society to engage more practically, because 'Some of the deficiencies in the GDC can be remedied in *how* it is implemented.'

Despite the weaknesses of the GDC, the UN Secretary-General then proposed to the General Assembly in November 2024 that a new Office for Digital and Emerging Technologies (ODET) should be created, still headed up by Gill, that would incorporate the existing activities of his Tech Envoy and strengthen system-wide coordination on digital cooperation to implement the GDC. The additional costs of this Office were originally budgeted at US$2.69 million for 2025 (revised to US$3.29 in paragraph 66d; UNGA, 2024) and US$3.44 million for 2026 and subsequent years, at a time when the UN's wider budget is under considerable stress (Mishra, 2025). While the efficacy of this new office remains to be seen, there seems little doubt that it represents a further effort by the Secretary-General and those close to him to centralise control over most matters relating to digital tech, and especially AI (see further in Chapter 5) within the UN system. As noted in Chapter 1, it is salient to recall here that the Secretary-General's AI Board, created in 2023, is mainly composed of people from scientific and corporate backgrounds with very little knowledge of the lived experiences of the world's poorest and most marginalised people. Much will depend on the quality of staff appointed to the new ODET, but at the very least this move is likely to disincentivise the ongoing work of competent people with real experience elsewhere within UN agencies in this broad field. It is surely no coincidence that the paper on the revised estimates resulting from the Pact for the Future (UNGA, 2024) noted above makes no mention of poverty, equity, or inequality, but does mention Artificial Intelligence 18 times.

The GDC was a lost opportunity to serve the interests of the world's poorest and most marginalised people. It did well, though, in serving the interests behind it in the UN Secretariat, some UN agencies, private sector companies, and certain powerful states and individuals. It was also a lost opportunity to guide international discourse on three of the most important aspects of digital tech use in the future: (1) the fundamental changes taking place in the relationships between machines and humans, as exemplified in our cyborg and robotic future (neither terms are mentioned at all in the GDC, although there are 35 mentions of AI); (2) the implications for human survival on Planet Earth as a result of the techno-digital-innovation complex that is coming to the fore and causing irreparable environmental damage (see also Chapter 5); and (3) the increasing loss of freedom (or enslavement) caused by the owners and designers of digital tech corporations (see Chapter 6). Soon after the publication of the final text of the GDC, the Office of the Tech

Envoy (now the UNODET) opened a process for its endorsement by governments and other entities (UN, 2024b), and one way through which those who wish to change the balance of power and serve the interests of the world's poorest and most marginalised is simply not to endorse it. Its acceptance by the UN General Assembly is, though, likely to mean that at some level it will continue to play a role in supporting the use of digital tech to create greater inequalities in the world. In contrast, refusing to be complicit in the UN Secretariat's self-serving agenda is an important step towards digital emancipation. It is perhaps salient that by the end of 2024 only 93 governments and 132 other organisations had actually endorsed the GDC (UN, 2024e; UN GDC, 2025); by November 2025, this had not been updated, indicating either general disinterest in the process, or a lack of attention to detail within ODET. As with so many UN initiatives, those organisations that did endorse the GDC have done so in large part because of hoped-for publicity for their own activities, rather than any deep commitment actually to delivering the content of the GDC. Saliently, few of the leading digital tech for development (or ICT4D) initiatives are visible among the endorsements.

Towards a new vision for the UN digital system

There remains a need for a global body that can bring states and their leaders together to deliver on the four original rationales for the existence of the UN cited in the epigraph to this chapter. There are also many very capable and well-intentioned people working within the UN system. Most of them aspire to 'do good' in their roles as best they can. However, as this chapter has highlighted, the UN system as a whole continually fails both them and the peoples its constituent parts claim to serve, because of its size and bureaucracy, its aspirations to try to do too much, the mission creep between agencies, the inappropriateness of some of its leaders, and the overall cost of its endeavours. The meaninglessness of the Summit of the Future suggests that there will be no significant rectifying change in the near future, but there are many relatively easy things that can be done to implement changes within the system that could indeed shift the balance in favour of the needs and interests of the world's poorest and most marginalised. In conclusion, this section offers five suggestions for actions that could readily be taken within the UN system that would begin to empower the most marginalised through the use of digital tech (for a more detailed analysis, see Unwin, 2023b and Chapter 7).

First, the central focus of the UN's digital agenda needs to change so that it is on equity, reducing inequalities, and mitigating the harms, both personal and environmental, caused by the design, building, and use of digital tech if the poorest and most marginalised are to have any hope of benefiting. This requires a fundamental shift in thinking by all those involved within the UN away from economic growth and the digital innovation agenda, especially their apparent fetishisation of Artificial Intelligence, and towards the principle of equity; economic growth must be balanced by equity.

Second, it is essential to reduce the amount of duplication, overlap, and mission creep within the UN system in the field of digital tech. To work effectively, this requires a much more hands-off approach from the central UN Secretariat, which should primarily focus on serving the heads of state of its member countries and the broad political agendas around peace and prosperity for which the UN was originally established, leaving the technical advice and delivery of digital tech to its funders, programmes, and specialist agencies. The creation of an Office for Digital and Emerging Technologies (ODET) seems to be a costly and harmful exercise, and the cutbacks to its budget in late 2025 are likely to make it even less able to deliver on its ambitious agendas. It is hard to see ODET delivering its claimed objectives with only six staff and a total of US$1,623,200 approved in the revised budget for 2025, rising to nine staff and US$1,898,500 for 2026 confirmed in September 2025 (UNGA, 2025). There was no need for the Secretary-General to create an entirely new apparatus around a Tech Envoy and the GDC, which has in practice primarily served to create a new power-base within the UN Secretariat through which the private sector can gain additional leverage over wider UN decision making. The UN has a perfectly respectable Chief Executives Board (CEB) where issues and policies across and between different agencies can be coordinated and delivered. Strengthening the work of the CEB would have been a much wiser and more effective approach than creating the ODET. Moreover, there also seems little need to convene so many overlapping UN digital forums and advisory boards as exist at present. This would, though, require those states, organisations, and individuals who have a vested interest in each of them to de-emphasise their own positions and work together in trust for the common good.

Third, UN agencies rather than the Secretariat should primarily be responsible for reaching inter-governmental agreements and providing high-quality advice and support on digital tech matters to governments. It is governments, not the UN Secretariat that have the key responsibility for delivering effective development for their citizens on the ground. Moreover, UN agencies should not in general be involved directly in delivering digital projects and programmes on the ground for which they do not have sufficient resources or competence. Where countries require external financial funding for the initiatives that they wish to implement, then they should be encouraged and supported in seeking such assistance from the existing donor communities, rather than from the UN, although the expertise present in agencies and programmes such as UNDP can often contribute effectively on how best to manage and use these resources in the interests of the poor.

The ITU should also retain its long-established responsibility for international agreements and standards relating to telecommunications and wireless, and would also seem to be a sensible place for leadership over the wider field of digital tech, albeit in collaboration with other agencies. There is little evidence that the present leadership and organisation of ODET, particularly with its latest budget reductions, have the capacity to ensure that the world's poorest and most marginalised will indeed benefit from the use of digital tech in the future.

Fourth, the UN system must free itself from the shackles of the Digital Barons and their coterie. This does not mean that the advice of tech companies and scientists should be rejected, far from it, but it does mean that the balance of influence in important decision-making forums should be more equitable between civil society and the private sector. The notion of multistakeholderism should be replaced by true multi-sector discussion and agreement, as argued earlier in this chapter. The dominance of the private sector in the present UN Secretary-General's initiatives relating to digital tech must be reversed, and the Secretariat must focus much more on being in the service of governments and thus all their citizens, rather than being a vehicle through which global digital capital enslaves them.

Finally, in this chapter, the UN system needs to focus on doing a few important things really well with respect to digital tech, rather than trying to do too much and fail in the attempt. This applies across the board, and is especially important, given the increasingly limited funding available from member states to support the UN's work. UN agencies must not continually rely on private funding in support of their agendas, because such funding always has strings attached. It is also essential that realistic plans are urgently put in place to deliver a clear and coherent replacement for the SDGs after 2030, and these should focus especially on outlining the precise responsibilities of governments, the private sector and civil society in delivering a just and equitable digital future for all (see further in Chapter 7).

Retreads: Pushing new rocks up new hills

Revi Sterling

(Found ICT4D in 2002; connecting women with information since 2003)

I recently read a gender digital divide/digital inclusion report. I was looking for some new example, case study, nugget of inclusionary gold. I imagine you know where this is leading. I did not find the gold. I found retreads. The report is glossy. It is a nice read if it is your first time with the topic. It is a pale comparison to many of the less-shiny but far richer knowledge products that preceded it. *Everything old is new again.* This is the leitmotif that plays in my head like an earworm. To torture myself, I compare the state of current offerings with the creative approaches we took with the Women-Connect Challenge and similar programs. Those did contain nuggets. Multitudes.

There is no doubt that digital inclusion is having a moment, but I wish I was more energized when I read that AI risks alienating rural women. I would like to be more grateful to the AI hype that is driving attention to digital divides. A new generation hears from GSMA that women's mobile internet use has stalled. Three billion offline is thrown around like a rallying cry. The World Economic Forum and McKinsey get to toss around dire GDP implications.

Instead, I wonder what we did wrong for the last 20 years. Were we ahead of ourselves? Were we just Cassandras warning people about restrictive social norms and technology determinism? It's a petty bitterness I have. There's a twinge of resentment when I look at the 'new' gender and digital inclusion research agendas – retreads if there ever were. No one has heard of the canon of digital divide research we built. If they have, they say 'It's so old!' but if you don't address a gaping wound, it doesn't heal.

I too, though, am a retread. I feel simultaneously forgotten and ahead of my time. In 2005, I thought I'd missed the boat – that all the great work had already been done by people like Hafkin and Huyer and Rathgeber and Adera and all the others; these women had already been at it for years. I have recently been asked to develop digital gender inclusion courses, host webinars, and create compelling videos that illustrate the benefits for women of being online. This would feel downright Sisyphean but for the fact that the present generation believes they have uncovered this gender injustice for the first time, even with a 20+ year starting base for them. It is too simple to say that 'people don't read'. I have come to believe that people do not want to be educated or to be historians. They want to be heroes. Just like I wanted to be.

What role for us, the retreads, then, especially as development systems shift dramatically? 'Development' now is not about progress; it's about keeping progress from slipping. Our goal is to hold the gendered rock. The goal of development work now may be maintaining a state – instead of expecting state change – until newer systems evolve to fill the vacuum. We need to be content, even dedicated, to holding the boulder on the steep hillside and not letting it slide. Everything old is new again, except new problems. This world promises us endless sources of new problems. Time to hold the rock and push some new rocks up new hills.

4 The 'me' syndrome

Tim Unwin

> [T]he fair's techno-optimism vision must have affected me. At that impressionable age, the message in 1962 was so clear: We would explore space, stop disease, travel faster and easier. Technology was progress, and in the right hands, it would bring peace.
>
> (Gates, 2025, p. 35)

> Move fast and break things . . . Unless you are breaking stuff . . . you are not moving fast enough.
>
> (Zuckerberg, cited in *Business Insider*, 2009)

> I have always believed that technology drives human civilization's endeavour and progress. So, if you ask what is my obsession, it is technology.
>
> (Ambani, cited in Singh *et al.*, 2016)

> I believe in empathy, like, I think you should care about other people . . . but you need to have empathy for, for civilisation as a whole, and not commit to a civilisational suicide . . . The fundamental weakness of Western civilisation is empathy.
>
> (Musk, 2025, *The Joe Rogan Experience*, 28 February 2025)

There is nothing 'natural' about the ways in which digital technologies have been developed. They are not somehow inexorable consequences of disembodied processes. They are not merely the products of wraithlike global corporations intent on profit and market share alone. They do not, as Ambani suggests above, drive human civilisation. There is likewise nothing inevitable about the application of these technologies in the delivery of international development initiatives. Digital technologies and the ways that they are utilised are the explicit results of their design, production, and use by *individual humans*, often coming together in teams and organisations with specific interests and intentions. It is the decisions and choices made by individuals such as those cited in the epigraphs above that ensure

DOI: 10.4324/9781003598008-4

that the world's poorest and most marginalised have been systematically more disadvantaged as a result of 'digitalisation'.

This chapter explores the motivations behind these individuals and the organisations within which they work, so as to understand better what must be changed if the world's poor are really to benefit from the use of digital tech. The quotation from Gates (2025) shares the enthusiasm and excitement that many of us felt in the 1960s and 1970s about the potential of digital tech. Zuckerberg's early comments introduce the importance of Agile, which is discussed further by Proctor later in this chapter (p. 90). Musk challenges us to rethink the grounds for empathy and altruism, which are a core focus for this chapter and much of Chapter 5.

An overwhelming focus on the 'me' or 'I' in digital tech, rather than 'we', has been one of the most pervasive factors influencing the increased inequalities caused by the design and use of digital tech. It is no coincidence that the term '*personal* computer' (PC) became universally used to refer to mass-market digital devices in the 1970s (Figure 4.1) (see Scullin, 2022). Calling these devices 'personal' implied that everyone could, or should, have one, in contrast to the older mainframe computers that were only affordable to large companies or governments. Everyone who was rich enough to buy one formed the market for these new smaller devices; the economically poor were still left behind.

This was something very different from the notion of 'communal digital devices'. In contrast, the telecentre movement of the late 1990s and 2000s was at least one attempt to retain some kind of collective digital presence that could be of benefit to many, rather than just the few (Ceballos *et al.*, 2006; Liyanage, 2009; see also vignettes by Rangaswamy, p. 131, and Tarmizi, p. 175). Nevertheless, the production and marketing of cheaper, smaller and more powerful mobile phones and tablets from the mid-2000s onwards rapidly reduced the economic viability of such telecentres. Powerful marketing encouraged consumers who could afford them to buy and regularly update these 'personal' devices, leading to a global smartphone market size of US$649.13 billion by 2024 (Skyquest, 2025). The poorest, those unable to purchase even a basic mobile device were left further behind, and global inequalities thus increased as a result of this individualisation. It is useful here to recall that about a quarter of the world's population still does not own a mobile phone in 2025.

An even more powerful and pervasive impact of individualism within the digital tech sector, particularly in the context of international development, has been the importance placed by the individuals involved in the sector on their own values and interests (see Note Two in Chapter 1, and the epigraphs to this chapter). This applies at all scales and contexts, from individual tech entrepreneurs to those working on 'digital for development' in small civil society organisations, to global corporations and international organisations advocating the use of digital tech to deliver the Sustainable Development Goals (SDGs). This chapter explores the factors that lie behind this emphasis on 'I' or 'me' rather than 'we', focusing mainly on the private sector, so as to uncover the complex mirage that has been constructed around the notion that digital tech creates a better world for all, including the

Figure 4.1 (a) The origins of 'personal computing' in Walt Kelly's Pogo cartoon of 19 May 1968; and (b) Hewlett Packard's advertisement of 4 October 1968 in *Science Magazine*

Source: (a) Walt Kelly, Fantagraphics, available at: https://hpmuseum.net/display_item.php?hw=50; (b) with grateful thanks to the HP Museum.

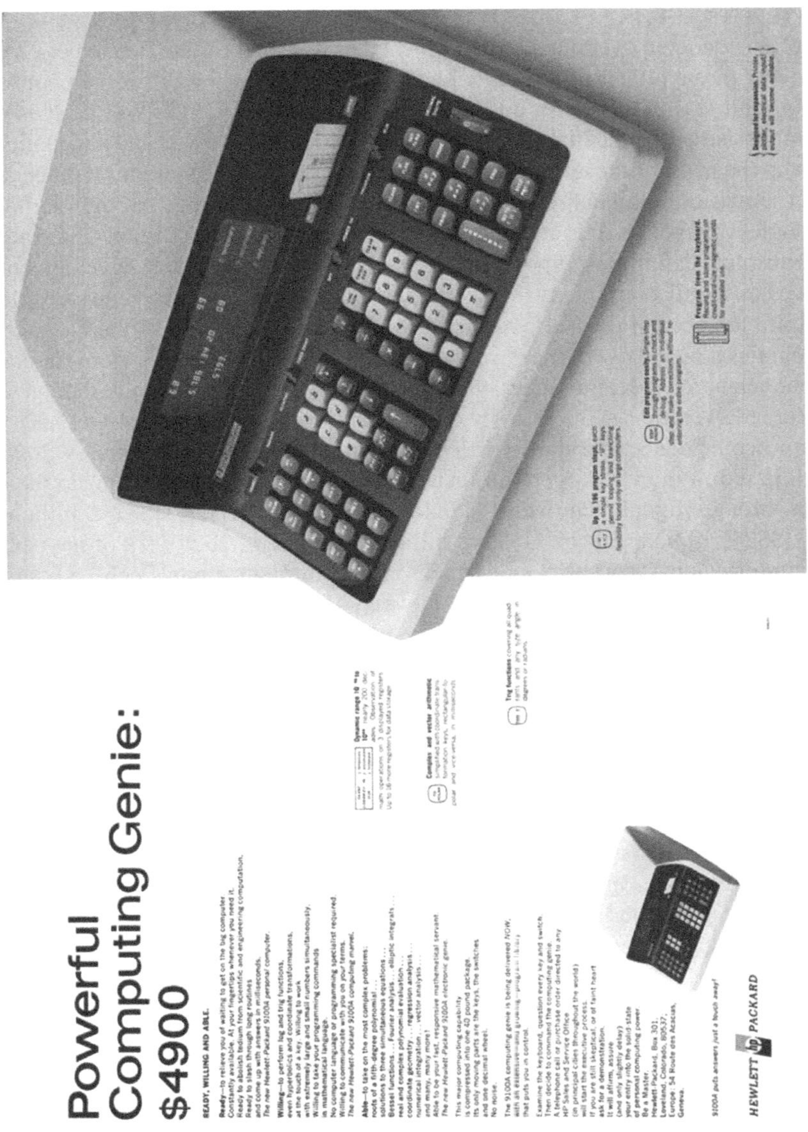

Figure 4.1 (Continued)

poorest and most marginalised (see vignettes by Nwakanma, p. 16, Harindranath, p. 96, and Juliette Unwin, p. 85).

Making money and doing good: are they compatible?

I have often pondered over the question of whether it is right for the rich to get very much richer through their use of digital tech if this is the only way for the poor to get a little bit less poor (Walsh and Lynch, 2008; Lobosco, 2012; Walsh, 2024). This emphasises the essential tension between economic growth and inequality that lies at the heart of the dominant global capitalist economic system. It embeds the excuse that rich philanthropists who have gained their wealth through digital technologies often offer. It is at the heart of the SDGs, the Global Digital Compact, and the absolute definitions of poverty that are so often used to justify development initiatives to 'eliminate poverty'. It is the inspiration for countless young people who aspire to successful careers as app developers and coders. Yet the question itself contains a fatal flaw. The answer is obvious. It lies in the simple word 'only'. The poor can surely become richer without the obscene levels of wealth that have been accrued over the last 30 years by the Digital Barons, and their acolytes.

It is no coincidence that the four richest people in the world in January 2025 were all based in the USA, were all men, were all white, and had all made their fortunes through digital tech (Musk US$402 billion, Zuckerberg US$248 billion, Bezos US$247 billion, Ellison US$248 billion; Bloomberg, 2025); six of the next ten, including Gates, had also made their wealth through digital tech. Oxfam (2025) notes that in 2024 the richest 1% owned 45% of the world's wealth, at a time when 44% of humanity lived below the World Bank poverty line of US$6.85 a day; 1 in 10 women live in extreme poverty (below US$2.15 a day). The wealthiest tech entrepreneurs and those who aspire to be like them (Seth, 2022) could use their expertise explicitly to address the needs of the world's poorest and most marginalised, and thus reduce inequalities. However, their choices and actions have instead led to them becoming extremely wealthy, while the world's poorest remain as relatively poor as they have always been (see also Chapter 6). The actions of most tech entrepreneurs have mainly served their own interests, the 'me', rather than the interests of the many, the 'we'.

One of the most important differences between humans and machines is humanity's rich diversity, especially our neurodiversity, the unique ways in which each of our brains work. When teaching computer scientists, especially those studying robotics, I often ask them 'would you build a robot with a disability?'. The answer has always been a unanimous 'of course not'. Yet it is this diversity that might in the end save the human race from its seemingly inevitable cyborg future (see also Chapter 6; Smith, 2025). We all live at different points on multiple spectra. Some of these are pathologised and/or medicalised (Sholl, 2017), and given legal meaning, as with particular types of disability, ethnic and gender differences, or health characteristics (Freeman, 2025; O'Sullivan, 2025). Frances (2013; 2025) goes further and argues that there has been out-of-control psychiatric diagnosis, especially around autism, through which too many healthy people are harmfully stigmatised as being mentally ill. However, many aspects of being human have not yet been classified by 'science' and made 'official', and it is therefore impossible to say definitively that a particular set of characteristics exists that might predispose some people to become successful digital entrepreneurs and immensely wealthy as a result. There is no one pathology that causes this condition.

The careers and biographies of the leading tech entrepreneurs nevertheless reveal some of the key characteristics that many of them had in common during their early lives: a fascination with science and the 'purity' of mathematics; strong determination and drive; an ability to shut other things out and focus on the task in hand; a thirst for personal success, defined mainly in terms of financial reward and fame; and most also came from fairly privileged or well-off backgrounds with supportive parents (Brandt, 2011; Arthur, 2012; Beahm, 2012; Isaacson, 2015; 2025; Vance, 2016; Schwab, 2023; 2025; Sharma *et al*., 2024; Gates, 2025). They are nevertheless all individuals with different personal circumstances and contexts: some are surprisingly shy, whereas others are brash and visibly arrogant; some have clearly identifiable neurodevelopmental characteristics, yet others appear not to. As Bezos, who has been diagnosed with ADHD, has said, there are a thousand ways to be smart (Fridman, 2023).

One thing that most of them share in common is that their focus has very much been on themselves and their interests, on 'me' rather than 'we', at least early in their careers before they became multi-billionaires. None of them got rich by putting the interests of the world's poorest and most marginalised at the heart of their business models. None of them directly experienced what it is like to be a poor African, Asian, or South American woman striving for life in a refugee camp in Africa, a favela in South America, or an isolated mountain village in Asia. Yet, across the world, the success of digital entrepreneurs such as these serves as a dream to millions of young people to gain computing skills, to try to design apps and AI systems, to become social media superstars (Hund, 2023; Hunsings and Senft, 2023), and one day to become rich enough to enjoy the luxurious trappings of digital power. Most of the very rich within the digital tech sector have put themselves and their ways of life first (Seth, 2022), and even when they have invested in philanthropy later in their lives, they have usually focused on enhancing continued economic growth which in turn has perpetuated structural inequalities. Two related aspects of this aspirational energy are particularly useful in seeking to understand why this has been so: neurodiversity and altruism.

Neurodiversity

Gates (2025) comments in his recent autobiography on the energy, like a jolt of electricity (p. 93) he got from the purity of maths and coding: 'To me, math was easy, even fun. I liked its ironclad certainty. Math followed basic rules; all you had to do was remember them' (p. 58), and 'I loved how the computer forced me to think. It was completely unforgiving in the face of mental sloppiness. It demanded that I be logically consistent and pay attention to details' (p. 94). There was indeed something pleasingly visceral about working with computers in the mid-1970s, not least experiencing their distinctive smells and the tactility of punch cards and tapes, and this combination of the 'physical' with intellectual rigour definitely appealed to many of those involved in computing at that time (Unwin, 2017a, pp. 1–2).

Second, though, many of the Digital Barons also consciously understood themselves from an early age as being different from the culturally and politically accepted behavioural norms of the societies in which they were living, or what is now often termed as being neurodivergent. In recent years 'neurodiversity' has come to be closely identified with one particular set of behavioural symptoms commonly associated with the autistic spectrum, although I prefer to retain the term for any set of behaviours that diverge from what a society sees as the norm. The autism spectrum has both positive and negative characteristics, and is often seen as one or more of the following characteristics: difficulties in social interaction, high anxiety levels, constant movement, sensitivity to external stimuli, and repeated routine behaviours (Fletcher-Watson and Happé, 2019). Alongside these, though, it must be stressed that many individuals identified as being on the autism spectrum also have particular strengths, often having high IQs, considerable creative or artistic ability, the capacity to focus for a long time on one thing, strong interests in technical fields, and great attention to detail (Frances, 2025). This combination of strengths is widely seen as being characteristic of many of those at the forefront of the digital tech sector in the twenty-first century.

Gates (2025) himself suggests that if he had been brought up in the 2020s he would probably have been diagnosed as being on the autism spectrum. Perhaps responses by those around him to such a diagnosis might actually have limited his abilities to be as successful as he has been. Zuckerberg stated in 2023 that he had Asperger's syndrome, now usually seen as being part of the autism spectrum. Bezos is reported to have been diagnosed with ADHD (Spies Coaching, 2023; Thomson, 2025), which frequently co-occurs with autism. Musk likewise reflected on having Asperger's on stage at TED2022, when he said that he 'would just tend to take things very literally . . . but then that turned out to be wrong' (King, 2022). He had previously spoken on the topic in 2021, saying, when hosting *Saturday Night Live*, that he 'found it rewarding to spend all night programming computers, just by myself. . . . But I think that is not normal', and also that it was certainly possible from a technology perspective that there was a link between his Asperger's and his companies (King, 2022). There has also been speculation that other leading figures in the tech sector are on the autism spectrum, but there are equally examples of those who have been successful with very different characteristics.

Given the neurodivergence of many of those involved in the digital sector, it can be conjectured that the ways through which they have shaped the characteristics of digital tech may themselves have enhanced neurodivergent characteristics among users, especially the young at particularly sensitive stages in their cognitive development. The addictive character of much digital tech (UK Addiction Treatment Centres, 2025) and the increased reporting of neurodivergence among young people, for example, might therefore both in part reflect the neurodivergent characteristics of those who have designed widely used digital technologies. This is a topic on which much more research is urgently required. Moreover, increasing evidence also suggests that using AI assistants can lead to an accumulation of cognitive debt (Kosmyna *et al.*, 2025). In a useful overview, the BiologyInsights Team (2025) emphasise:

> The developing brain relies on sensory input to shape and refine its neural architecture. From birth, children engage with their environment through vision, touch, hearing, taste, and smell, each modality contributing to specialised neural circuits. Sensory experiences drive plasticity by reinforcing frequently stimulated pathways while diminishing inactive ones.

If there is an early overload of only some of the senses as a result of using digital tech, this is likely to have a lasting impact on subsequent brain development, although it should also be recognised that there is growing evidence that some degree of brain plasticity can remain throughout life (Dennis *et al.*, 2013; Pauwels *et al.*, 2018) and therefore these impacts may be reversible.

Interestingly, attempts to build companies around the specific skills of individuals with autism have not been particularly successful. Bernick (2024), for example, has traced the wave of tech start-ups in the 2010s that arose specifically 'with the goal of employing adults with autism, building on the claimed skills of these adults in software testing and quality control', and concluded that most have failed. A few, such as AutonomyWorks, Ultranauts, and VenturesATL, have nevertheless survived, mainly because of the particular skills sets of their founding entrepreneurs who were able to pivot to new markets (see also Autus, 2025; Lexxic, 2024).

Subsequent altruism

There is little doubt that the digital tech sector has mostly been driven by people whose prime driver has been their fascination with the potential of digital technologies, and an interest in using this to build their own wealth and power.

It has frequently been all about 'Me'. While this has usually been focused mainly on economic power, the recent turmoil following President Trump's election to a second term of office from January 2025 illustrates clearly that many also have political interests, if only to protect their economic power and wealth. As Ahmed (2025) notes, the attendance of big tech CEOs at Trump's inauguration signalled 'a profound change in how corporate America aligns itself with political power, raising pressing questions about the influence of technology conglomerates on governance and democracy'. At the extreme, this was well illustrated in the tumultuous relationship between Musk and Trump following the latter's election (Bedigan, 2024), which highlighted Musk's short-term ability to wield both great economic and political power.

As Musk's erstwhile colleague Low (2025) commented:

> Elon believes he is above everyone else. He used to think he worked on the most important problems. When I met him, he did not presume to be a technical person – he would be the first to say that he lacked the expertise to

> understand certain data. That happened later. Now, he believes he has all the solutions . . . Everything Elon does is about acquiring and consolidating power.

Altruism was not at the heart of what Musk and most other tech billionaires were doing when they were building their digital tech empires in the first quarter of the twenty-first century.

The original notion of *altruism* as a concern for others completely independent of any self-interest or reciprocity, was used by Comte (1875) in the mid-nineteenth century as an antonym for *egoism*, the role of the self as the motivation and goal of a person's actions. Both terms have been subject to considerable philosophical debate in subsequent years (Rand, 1964; O'Neill, 1996; 2016; Reginster, 2000; Nietzsche, 2008). There is, though, general recognition that they serve as idealised extremes at the ends of one of the many spectra of human behaviours with respect to others. Within such a framing, most of those involved in digital tech companies are much closer to the egoistic end of the spectrum than they are to the altruistic. Nevertheless, there are exceptions, especially among less personally ambitious entrepreneurs (Banks, 2013a; 2022) (see vignettes by Aggarwal, p. 81, Banks, p. 100, and Hughes, p. 115), and a few large digital corporations have also sought to adopt a more balanced approach. Cisco, for example, founded in 1984 by a group of computer scientists at Stanford, placed corporate social responsibility (CSR) alongside customer focus as its two core values from the very beginning (Cisco, 2004). This has been particularly evident through the work of the Cisco Foundation, created in 1997, which has provided substantial levels of funding for civil society organisations. Even today, in a much more competitive commercial context, Cisco continues to prioritise its 'commitment to positively impact the world around us' (Cisco, 2024, p. 5). Chambers (2018), who was CEO of Cisco from 1995–2015, in many ways embodied the ethos of the company at that time, and stands in marked contrast to the extrovert egoism of many of his peers. Diagnosed with dyslexia at the age of 9, he could be nervous about speaking on stage, and he often spoke of his deep concern for ensuring that the company did indeed deliver on its CSR agenda. This was catalysed and delivered in practice by Tae Yoo who had joined Cisco in 1989 and subsequently became senior vice president of corporate affairs and sustainability there (Morris and Yoo, 2022). Likewise, Jack Ma, the founder of Alibaba in China in 1999, has always believed in the importance of successful businesses not only in making profits, but also in creating wider value for society. He came from a humble background, was not good at maths in school, and did not have an elite education, but became a teacher and then identified an opportunity to develop Alibaba as a company designed especially to help small businesses to leverage innovation and technology so that they could grow in domestic and external markets. His subsequent creation of the Ali Baba and Jack Ma Foundations, focusing on a range of interests including education, the environment, and women's leadership emphasise his long-standing commitment to altruism (see also China Development Brief, n.d.; Jack Ma Foundation, n.d.).

These examples stand in marked contrast to most of the purportedly philanthropic initiatives and charitable foundations created by other digital corporations and their founders and leaders (see Note Three in Chapter 1; Maclean *et al*., 2021). Most such attempts to combine wealth with doing good have ultimately failed. One of the most high profile of these notionally altruistic initiatives was Internet.org, launched by Zuckerberg as a top priority within Facebook in 2013. This initially had some potential to enable many of the world's poor and marginalised peoples to improve their lives by connecting to the Internet. Hempel (2018) thus notes that it was originally pitched as a humanitarian effort, and that Zuckerberg claimed he wasn't expecting a profit from it, 'suggesting he was motivated by a moral imperative, not a financial one'. However, in developing Internet.org Zuckerberg and Facebook chose not to learn from the experience of many of those who had previously sought to deliver digital tech for development initiatives across the world. Very quickly, Internet.org was criticised for trying to create a two-tiered Internet and in 2015 morphed into Free Basics. Even then, India subsequently decided in 2016 to block Free Basics on the grounds that it was breaking the principles of net neutrality. Facebook's Aquila project to create solar-powered drones to deliver the Internet was likewise closed in 2018, following only two flights in 2016 and 2017. Hempel (2018) gives Zuckerberg the benefit of the doubt, suggesting that:

> he was sincere in his swashbuckling belief that Facebook was among a small number of players that had the money, know-how, and global reach to fast-forward history, jump-starting the economic lives of the 5 billion people who do not yet surf the web.

Nevertheless, one of Zuckerberg's great strengths is his ability to see, and create, commercial opportunities from which he and Meta can benefit (Kelleher, 2013). Internet.org was just one way in which he was seeking at that time to get as many people as possible to use his platforms, especially in poorer parts of the world that had not yet been exploited by other digital tech entrepreneurs, so that he could eventually monetise their engagement.

Facebook's efforts with Internet.org to connect the world have not been the only purportedly philanthropic high-profile infrastructure failures. Google's similarly ambitious Loon project launched in 2011 to deliver Internet connectivity through a balloon network was likewise closed by Alphabet in 2021 because of a lack of profitability; its Project Titan which had been created following Google's purchase of Titan Aerospace in 2014 to use drones for similar purposes had been closed earlier in 2017.

There have been numerous criticisms of the self-interest of the supposedly philanthropic and altruistic initiatives of many of the wealthiest tech entrepreneurs (Edwards, 2009; Rushton and Williams, 2011; Curtis, 2016). Schwab (2023) has in particular developed a comprehensive critique of the Bill and Melinda Gates Foundation and the significant increase in Gates' own personal investments since leaving day-to-day operations at Microsoft in 2008 (and stepping down as Chairman in 2014). He provides a powerful indictment of the failure of philanthropy resulting from digital tech to be used wisely in the interests of the world's poorest (see also *Forbes*, 2025; Razzall, 2025). Gates' investments in vaccines and healthcare have undoubtedly saved the lives of many people, especially in the economically poorer countries of the world, but he has also made significant profits from them. There are four main problems with Gates' approach to international development, and specifically with the role of digital tech therein. First, he really does believe that his is the right, and perhaps only valid approach to delivering effective development, primarily because this is based on his own success. His former wife recalls that Gates 'may not have known everything, but he sure gives that impression' (Aitkenhead, 2025, p. 13). Unfortunately, he has rather little direct experience of what has shaped the lives of the world's poorest and most marginalised peoples, and focuses primarily on increasing economic growth rather than reducing inequalities. Second, his original wealth came from the profits of Microsoft, which were built on generating revenue from the sale of software, services, and hardware. If he, and his leadership team, had been more interested in equity rather than profit, they could have used their undoubted skills to create an entirely different business model that could have begun to redress the equity balance between the world's richest and poorest. He chose instead to make himself and his company very rich. Third, his Foundation reports only to its Governing Body of which he is chair and it lacks any wider form of democratic representation and responsibility (Schwab, 2023). Especially in its early days, the Foundation explicitly remained apart from the policies and practices of traditional bilateral donors and thus distorted the existing global system of financing international development through institutions such as the OECD's Development Assistance Committee (Edwards, 2009; Curtis, 2016). In recent years, the Foundation has certainly worked much more closely with donors, such as USAID, the World Bank, and the UK's DFID (now FCDO), but frequently it still tries to influence them into its own ways of approaching 'development', rather than acceding to what it tends to see as the flawed traditional ODA system (see also Malkan, 2025, for a more recent critique of the Foundation's work in trying to support African farmers).

The authoritarian power of the Foundation is all too visible to those who have worked with it, and has been well documented by Schwab (2023). Finally, it is remarkable for someone who has made his career and wealth out of digital tech, that he has only rarely chosen to support the wise use of digital for development through his Foundation, although Microsoft has indeed often partnered with other organisations to deliver such interventions, thereby contributing to substantial increases in the sales of its products.

Gates is an extreme example, but much of the above critique also applies to other foundations established by tech entrepreneurs. Most of them are as much about raising the profiles of their founders, as they are concerned with using digital tech to serve the interests of the world's poorest. Importantly, most are also driven by the same agendas that made their founders rich in the first place, and see 'development' as being concerned primarily with increasing the economic growth that will ultimately also benefit their former companies rather than specifically reducing global inequalities (Devex, 2023). There are, though, a few important exceptions to this generalisation. Scott, Bezos' former wife, for example, has adopted a notably different 'hands off' and 'no strings attached' approach to tackle equity issues through her foundation, although this focuses mainly on issues within the USA. Likewise, the Oak Foundation, one of largest 10 foundations contributing to international development, although not directly deriving its wealth from the digital tech sector, has an overt focus on justice, equity, diversity, and inclusion.

With increasing ambition to try to show that altruism does indeed deliver positive outcomes, several tech entrepreneurs, most notably Moskovitz, Musk, and the since-disgraced Bankman-Fried, have recently begun to champion 'effective altruism', a philosophy and practice developed in the early twenty-first century by a small group of philosophers at Oxford University, notably Ord (2014; 2020) and MacAskill (2016), building on Singer's (1979; 2002; 2015) ethical decision-making framework. In 2011, the Centre for Effective Altruism was founded as an umbrella organisation for two earlier entities, Giving What We Can and 80,000 Hours, to advance this agenda, focusing especially on identifying optimal ways to deliver solutions to the world's most pressing problems based on four main principles: prioritisation, impartial altruism, open truth-seeking, and a collaborative spirit (Effective Altruism, n.d.). This has provoked much debate and critiques from different perspectives, both on philosophical and practical grounds (Clough, 2015; Crary, 2021), but has proved highly popular to many in the tech sector, in part because of its overt focus on trying to create solutions that balance both the potential for good of AI alongside the harms for which it can be used.

Much work in effective altruism is nevertheless problematic for the same reasons that Gates' efforts to make the world a better place remain flawed: they concentrate mainly on providing benefits 'to' the world's poor, they fail to tackle the underlying structures that cause inequalities, and they primarily reinforce the dominant economic growth-based model of 'development'. Although they recognise extreme inequalities, they see the answer as being primarily in identifying the most cost-effective charitable giving directly to the world's poor as summarised in the following quotation:

> Global inequality is extreme. Because of this, transferring resources to the very poorest people in the world can do a huge amount of good. In richer countries like the US and UK, governments are typically willing to spend over $1 million to save a life . . . This is well worth doing, but in the world's poorest countries, the cost of saving a life is far lower.
>
> (Effective Altruism, n.d.)

This has given rise to initiatives such as GiveWell, and the tech company Wave, which has enabled migrants to transfer remittances more cost effectively than heretofore. The latter can also be seen as one of many examples of Public Interest Technologies (PIT) that have largely been promoted since the early 2000s by organisations, especially in the USA, such as Code for America (2018), the Ford Foundation, Hack4Impact, and the Mozilla Foundation. Most such initiatives, though, are primarily focused on redistribution of benefits within the system, rather than fundamentally changing the system so that its benefits are redistributed more equitably by design. Moreover, as discussed further, relying on the largesse of the rich can be degrading and will never ensure sustainability. Effective altruism is elitist and based on the notion of finding the best ways to help others, but fails sufficiently to recognise that what its practitioners, the rich 'me', think are the most effective ways, may not actually be what the world's poorest and most marginalised want or need. Indeed, rather than breaking the shackles of digital enslavement, it can instead reinforce them.

It is important also to note that the 'me-syndrome' is not just a private sector characteristic, and applies in many other high-profile civil society and academic interventions that seek to promote digital tech for development (Unwin *et al.*, 2023). This is typified by the very different styles of Negroponte with his One Laptop per Child initiative (Ames, 2019; OLPC, 2023), and Sachs (2018), with his Millennium Villages (Ericsson, n.d.; Millennium Village Project, n.d.), both of which were launched in 2005. Both individuals are very able, with immense drive and high confidence in their own abilities, but their experiences also reveal that an emphasis on 'me' often does not in the end have positive outcomes when technology from the outside is used to try to improve the lives of the poorest and most marginalised, without sufficient understanding of the development context on the ground. Ames (2019) has thus shown how despite Negroponte's undoubted charisma and enthusiasm, OLPC fell short in many ways, partly reflecting his overconfidence but also the allure of technology hype for those who are eager to accept it without sufficient critical reflection. She further emphasises the harms that can result when utopian dreams from outsiders drive the use of digital tech in development contexts. Negroponte has nevertheless learned from his mistakes, and in recent years has adopted a much more nuanced and realistic approach to the use of digital tech for development.

Munk (2014) has likewise portrayed in considerable detail the great difficulties in delivering what Sachs thought should have positive impacts on the lives of rural Africans, despite his undoubted ability to bring together a diverse coalition of partners, especially from donors and the private sector to provide funding for what they saw as a 'good cause'. Sachs, with his 'preternaturally brilliant mind' (Munk, 2014, p. 9) is often described as being driven, messianic, and quixotic, but none of this helped him to deliver lasting improvements in the lives of those in the villages, let alone substantial numbers of the poorest and most marginalised elsewhere in Africa (see also Green, 2013). His overconfidence and belief in his abilities as an economist prevented him from sufficiently understanding the real challenges and needs of poor communities in rural Africa for the use of digital tech (Sachs, 2008). He would have done well to have learnt from previous initiatives that had also sought to use digital tech to improve the lives of the poor elsewhere in the world.

It's about who built it – and who didn't

Shabnam Aggarwal

(CTO, edtech innovator, AI strategist, failure evangelist, relentless kids advocate)

I've spent nearly two decades trying to build technology that changes the game for education. I've watched wave after wave crash and recede; each one launched with the language of transformation, while most quietly fizzled out.

In early 2025, I found myself in a familiar setting: a hot, crowded classroom in Rwanda, watching a teacher named Jean-Luc hover his phone over a student's exam script. We were prototyping LearnLens, our new AI-powered assessment grading tool (Rising Academies, 2025). 'How long will this take?' he asked sceptically, clearly wanting to get back to his students. When the class results popped up in his WhatsApp chat in seconds, his eyebrows shot up. 'Is that really it!?' That's usually the part in the story where someone wants me to say, 'See? The tech works!'. But here's the thing: it's not about the *tech*. It never was.

Jean-Luc wasn't just a user. He was part of the reason LearnLens exists in the first place. His and his colleagues' frustration about assessment marking tedium, their WhatsApp voicenotes on feature requests and language support, their insistence that if something didn't save them time, they weren't touching it (which I loved) – that's what shaped this tool. Jean-Luc didn't 'adopt' our solution. He built it with us.

This is the only real inflection point we're standing on in 2025. Not the AI. Not the large language models. Not the imminent promises of scale and cost efficacy and Artificial General Intelligence (AGI). But this: the realization that anything built for communities without being built *with* them will either fail, or worse – leave deep wounds.

We've done the 'Here's your shiny tool, now off you go!' routine so many times before. We've paraded into schools with tablets and dashboards and overconfidence. We've learned the hard way that technology doesn't empower people. *People empower technology*. And if the people at the centre – teachers, students, caregivers, school leaders – aren't part of the act of creation, the tool will eventually gather dust, or quietly erode the very trust and value it was supposed to build.

We're definitely not fully there yet. There's still a lot of saviourism, a lot of packaging of 'agency' as a feature instead of a process. But this time, I think we have a shot. If we stay honest, if we stay humble, and if we build with – not just *for* – the people we claim to serve, we might just avoid making the same mistakes again.

Self-critique and the overconfidence effect

The previous section has emphasised the importance that the personal characteristics of individuals have in creating the type of digital tech sector that currently exists. The sector has largely been designed in their own images, through their own lenses, and in their own interests. Some of these characteristics have also played an important part in obfuscating the truth about the efficacy of digital tech interventions in poorer areas of the world. It is, thus, remarkable how little self-critique there has been, despite the widespread and long-established recognition of the importance of effective and appropriate monitoring and evaluation (M&E) to enable those delivering such interventions to become learning organisations and deliver outcomes with the poor and marginalised that really benefit them (Wagner *et al.*, 2005; Lennie and Tacchi, 2013; Hollow, 2015; WHO, 2016; see the vignette by Aggarwal, p. 81). This section explores why those leading many such digital tech for development interventions seem to believe that there is no need for such exercises and that it is perfectly acceptable to 'build the plane while flying it'. The lack of effective self-critique and evaluation focusing on outcomes is one of the key reasons why digital tech continues to perpetuate inequalities. Two related issues help in understanding why so many digital tech for development interventions fail to deliver on the needs of the world's poorest.

Overconfidence in the digital tech for development context

The first reason is the widespread apparent overconfidence among many of those leading digital tech interventions, even in the development sector where it might be thought that there would be more understanding and empathy with the needs of the world's poorest. This represents a classic instance of the Dunning-Kruger effect, whereby 'people who are unskilled in these domains suffer a dual burden: not only do they reach erroneous conclusions and make unfortunate choices, but their incompetence robs them of the metacognitive ability to realize it' (Kruger and Dunning, 1999, p. 1121). Moore and Healy (2008) have usefully differentiated three ways in which overconfidence has been explored: (1) as overestimation of performance, (2) as overplacement of performance relative to others, and (3) as excessive precision in one's belief. Many, although certainly not all, of those involved in digital tech interventions exhibit aspects of all three of these, and yet as Plous (1993, p. 217) has commented, 'No problem in judgement and decision making is more prevalent and more potentially catastrophic than overconfidence.' This lack of willingness to explore previous digital tech initiatives, to learn from their mistakes, and to focus on improving them in the interests of the poorest is indeed shocking and surprising. As discussed in Note Two in Chapter 1, there are many reasons why such overconfidence exists, even if underlying it there may sometimes also be a deep sense of insecurity that drives a search for success. Five factors are particularly important in digital tech initiatives and especially those driven by practitioners from North America and Europe:

1. the widespread grade inflation in education systems, which gives many people near the beginning of their careers an inflated sense of their own real abilities (Chowdhury, 2018);
2. the introduction of private sector, profit-driven processes and mentalities in civil society and international organisations that often reward apparent immediate success over long-term change (McKinsey, 2020);

3. a shift in the workplace balance towards over-emphasis on praise rather than constructive criticism (Zenger and Folkman, 2013);
4. a belief in American exceptionalism (Lipset, 1997; O'Connor *et al.*, 2022) alongside an elitist view of the world that makes some unqualified people really believe that they do know best;
5. the scientistic attitude that shapes many people's belief that new digital technologies are inherently good (see also the discussion of innovation in Chapter 5).

As Asgard (2025) has commented, 'There is an insatiable desire among Gen Z to be impressive' that has given rise to 'the world of posturing'. This expression of the 'me syndrome' prevents many people from learning from the mistakes of others.

Monitoring, evaluation, and a failure to understand the most marginalised

There are also at least six very practical reasons why insufficient good quality M&E is undertaken within digital tech for development interventions, and these too have serious negative impacts on their ability to serve the interests of the world's poorest and most marginalised. First, effective and rigorous M&E is difficult and time-consuming. It should never be just an 'add-on' or afterthought merely to satisfy funders, but should instead be deeply embedded within the culture of any organisation committed to improving effectiveness and efficiency. For this reason, the term MERL (Monitoring, Evaluation, Research, and Learning) has recently become more widely used to describe the process (Tilton *et al.*, 2020). Regardless of the name, rigorous M&E is hard to do, and there remain remarkably few good practitioners in the field.

Second, high-quality M&E is therefore expensive, and when practitioners want to maximise their funds to deliver action on the ground, it is an easy temptation to cut back on M&E. Estimates of the amount that should be allocated to monitoring and evaluation vary widely, depending on the type of interventions being reviewed, but organisations such as UNDP, Oxfam, and Save the Children recommend that this figure should be between 3% and 5% of total programme budgets (EvalCommunity, n.d.). Others estimate that it should in some circumstances be much higher at 20–25% (Newman *et al.*, 2013). The reality is that during the 2010s, as a KPMG (2014) report suggested, many practitioners considered that less than 2% was actually being spent on M&E, and little has changed since then. My own experience suggests that as a good rule-of-thumb expenditure of around 10% on M&E is necessary if organisations are really to improve their practices in digital tech for development programmes, particularly if they are also to consider reviewing the unintended consequences (Unwin, 2020a).

Third, there is an overwhelming emphasis in many tech for development projects on choosing the easiest contexts in which to 'prove' their success, the so-called 'low-hanging fruit'. By definition, these are not going to show whether or not an initiative will deliver positive outcomes in the most disadvantaged and marginal areas.

Fourth, most such interventions never evaluate like for like. Thus, most digital tech interventions in education that do incorporate M&E within them compare results between the improved learning achievements in schools that have the intervention, with the results of those without the new learning opportunities provided through the digital intervention. This is a false comparison, because most interventions providing additional inputs of any kind will usually lead to improved results since teachers and learners involved feel special and respond positively. Comparisons are only valid if the digital intervention is compared with another intervention costing the same amount of money but doing differing things. Few organisations have the budget to experiment in such ways, especially in resource-poor contexts.

Fifth, far too many digital tech evaluations focus on inputs rather outcomes, only measuring such things as the numbers of digital devices introduced, the number of people using them, or the number of training sessions run. In large part, this is because of the physical reality of digital devices; it is easy to measure such inputs, but very much harder to measure the real outcomes of any intervention using them that may take a long time to come to fruition (see vignette by Hollow, p. 126). Moreover, those delivering such interventions frequently believe simply in the inevitable good of digital tech, and that by reporting the amount of tech that has been introduced they consider that they are providing an accurate measure of positive benefit.

Finally, in this list, there is also often a tendency for practitioners and academics alike to value most the evidence of more recent research, and thus fail to draw on established understandings of what works in practice from decades previously. This is interestingly reinforced by the character of digital tech itself, with the use of search algorithms frequently tending to emphasise the latest publications, which are not necessarily the best. Thousands of interesting and useful reports, books, and articles have been published at great cost on the successes and pitfalls of digital tech interventions in development over the last 25 years (see, for example, these very different types of publication: Beardon *et al.*, 2004; Maclean, 2004; Weigel and Waldburger, 2004; Trucano, 2005; Flickenger, 2006; Unwin, 2009; GIZ, 2012; Gyan, 2016) but most of these remain unknown and unread. There is thus very wasteful reinvention of the wheel by new individuals and organisations wishing to make a career and name for themselves in the field by duplicating such work without necessarily realising it. Likewise, most of the international ICT4D or Digital Tech for Development conferences present little that is not known or really new (see also discussion of UN summits in Chapter 3).

These problems are also very evident in the academic discourse around digital tech and development, where many researchers wishing to promote their own ideas as being novel fail sufficiently to explore older literatures that have made very similar arguments previously. In part, this lack of understanding of the genesis of ideas is a result of academic publishing silos and poor scholarship, with many authors insufficiently exploring the material previously published in other disciplines. It is also in part driven by rapid expansion of the academic publishing industry which often appears to show little regard for the rigour and novelty of the research published (Unwin, 2011; Tennant and Ross-Hellauer, 2020; Allen *et al.*, 2022; Riding, 2022). Most science is not the pure and definitive source of knowledge that it is often taken to be. Academic research serves the interests and biases of its practitioners exactly in the same way as it does in other professions (although see vignette by Juliette Unwin, p. 85 for an alternative approach).

All these factors combined have reinforced the largely self-promotional, rather than self-critical, approach that many people and organisations have taken towards the creation of digital tech interventions, and have failed sufficiently to understand how the poorest and most marginalised have often been excluded. The next section provides illustrations of the ways that this occurs not only at an individual level, but also at an institutional level within international organisations, companies, and civil society entities.

Crafting a more equitable framework for global epidemiological research practice: Working *with* not *on*

Juliette Unwin

(Using statistics and mathematical modelling to improve global public health)

Digital tech can indeed be used for good in the health sector, but this hasn't always been the case. As an emerging academic, I use other people's health data in my research on modelling infectious diseases (Unwin, H.J.T. *et al.*, 2022; 2023), including aggregated numbers of people infected by Ebola during an outbreak, childhood malaria prevalence from household surveys, and more recently electronic health records. I have seen first-hand how important it is to co-create my research with people in Africa and Asia if I want to consider disease transmission outside my own country. I can bring maths to the table, but I don't fully understand the intricacies of how the data have been collected or what interventions make the most sense in their context.

Historically, academics, many I'm sure unintentionally, have made their careers by using health-related data provided by other people and have not fully recognised the efforts of those who collected it. Having come from a background in engineering where I didn't really work with large data sets, this was something I struggled with in my earliest 'first author' publication. I was developing a new method and applied it to some data that had kindly been shared with me. But when it came to publication, the data owners had changed and couldn't be contacted, and we were a bit stuck. From that moment I decided I didn't want to end up in that position ever again.

The digital analysis of good quality health data can save lives. For example, powerful computers have been used during disease outbreaks to predict where diseases will spread and help plan who to vaccinate and where extra medical interventions are necessary. And – whatever your feelings about how the COVID-19 pandemic was managed – it would have been much worse if there hadn't been easy sharing of information around the globe.

Since the global burden of infectious disease is not equitable, it's important to help train my future colleagues around the world in methods and tools so they can model transmission themselves. Teaching courses and running workshops at Malawi Liverpool Wellcome in Blantyre in Malawi, the National Institute of Public Health in Cambodia, and the African Institute for Mathematical Sciences (AIMS) in Rwanda and Cape Town, has not only been personally fulfilling, but it has hopefully also helped lay the foundations for a more mutually beneficial and equitable model of impactful science.

The institutional 'me'

It is helpful to distinguish between three types of individuals and organisations because how to change their attitudes towards using digital tech in the interests of the poorest and most marginalised will differ accordingly: (1) those who are well intentioned with respect to international development but not doing good; (2) those who are well intentioned and doing good; and (3) those who do not really intend to contribute positively to development at all. The focus in this book is primarily on the first of these groups. How can they do the good that they hope to do? As outlined in Chapter 2, doing good in this context means contributing positively on equity and with the most marginalised, rather than merely just addressing the economic growth agenda. It is also crucial to differentiate between publicly stated intentions and underlying motivations. Many digital tech companies, for example, are only too pleased to sign up to globally agreed agendas, such as the UN's SDGs, or UNICEF and the ITU's Giga initiative, or the ITU's Partner2Connect coalition (see Chapter 3), because they want to appear to be contributing to something generally accepted as being worthy and 'good'. However, their underlying motivation is often that they see this as being an opportunity to expand their markets both directly and indirectly, to gain valuable networking opportunities, and to have a seat at international decision-making bodies, such as UNESCO and the ITU's Broadband Commission, or one of the UN Secretary-General's High Level digital panels, as outlined in Chapter 3. These are all mainly about 'me', both at a corporate and an individual level.

The private sector 'me'

Private sector companies are ultimately concerned with competing to make profits for their owners and shareholders. There is nothing wrong with this. Despite Prahalad and Hart's (2002) notion of a fortune at the bottom of the pyramid (BoP) (see also Prahalad, 2004) which became popular in the 2000s, it has proved difficult to achieve corporate success by focusing on a development perspective that addresses equity more than growth (Simanis, 2012) (although see vignette by Hughes, p. 115). Reliance's disruptive Jio telecommunication company, established in 2015 and with 40% of the Indian mobile market at the end of 2024, may be an exception, but even then it gained much of its early market share through dramatically undercutting the prices charged by other operators in the Indian market, a move that was criticised by incumbent operators as being supported by the Indian government and regulator, not least with soft loans (Parsheera and Trehan, 2022). In general terms, digital tech companies have very much focused on their own specific identities (the corporate 'me') in promoting their development credentials, and these emphasise the economic growth agenda very much more than they do equity. Sometimes this has involved working successfully in creative partnerships, as with Cisco, Intel and Microsoft in the World Economic Forum's Global Education Initiative between 2003 and 2011 (Unwin and Wong, 2012), or Imfundo's partnership of 40 organisations working together to deliver digital interventions in African education systems in the early 2000s (see Chapter 1), where each organisations brought their own particular resources and skills sets together for the greater good. However, more often than not, the creation and maintenance of a high and positive profile for a company and its founder(s) are one of the main factors in ensuring its competitive success. This is typified through sponsorship and speaking slots at international conferences and events, as well as supporting civil society entities to deliver practical interventions through their corporate social and environmental responsibility (CSER) departments, once again seeking to do good while also raising their corporate profile and market share. Initiatives supporting marginalised communities that are funded through CSER grants are nevertheless almost always palliative and short-term, because such funding is rarely sustainable in the longer term (see section on projectitis).

At a smaller scale, the 'me' syndrome is writ large in the entrepreneurial pitches required to gain funding for digital start-ups across the world (see Chapter 5). This applies as much to the early years of companies such as Facebook (now Meta) (PitchDeckHunt, n.d.) as it does to the pitches by young African entrepreneurs working in Silicon Savannah in the 2020s (Wahome, 2023). The personality of the entrepreneur, and their ability to convince angel investors, are crucial to success. Nevertheless, the motivations of those involved vary. Frequently, successful young entrepreneurs get caught up in a cycle of bidding for new funding, before they have actually produced anything sustainable from their first round of funding. Even if the original motivation was to do good by serving the interests of the poorest, this gets swept away as the cycle of growth comes to dominate in their efforts to provide a return on investment for their investors. Frequently, across Africa, Asia, and South America, the most successful young entrepreneurs are then made offers they find difficult to refuse, and are either bought out or offered positions in global corporations that take them away from their original interests and homelands.

The international agency 'me'

The 'me' syndrome also operates prominently within international organisations and the UN system. At the highest level this can be seen in the UN Secretary-General Guterres' personal agenda round the Summit of the Future and the GDC (see Chapter 3). However, it also operates at every level within UN agencies. People need to create highly visible and successful profiles that will be supported by their organisation's members. One way in which this is most effectively done is through launching new digital initiatives broadly aligned with global priorities around Agenda 2030, or within the priorities of individual agencies. Aspirant leaders seek to deliver digital for all, as defined and aligned within broader UN agendas, but ultimately their actions largely serve the growth agenda and interests of global companies rather than the world's poorest. This is particularly so when they seek to develop high profile partnerships with the private sector. Many novel UN initiatives are also grounded in a desire to attract further funding into the agency from the private sector and donor governments. WHO (2021), for example, launched a digital health strategy 2020–2025, with a key element including a desire to attract funding from donors and mHealth companies to deliver it. Likewise, the ITU's Partner2Connect initiative prided itself in having received commitments of US$73 billion by June 2025 to provide universal meaningful connectivity. More generally, the increasing proliferation of digital technologies across most development initiatives has had a tendency for competition between agencies (see Chapter 3), and those that have been successful have had high profile and diplomatically capable leaders who are able successfully to steer their 'institutional me', both through judicious collaboration with other agencies and where necessary through strong defence of their remits and responsibilities, as determined by their member governments.

In the 2010s and 2020s, some governments that had not traditionally funded digital tech for development initiatives (those not in the OECD's Development Assistance Committee, DAC; see Mawdsley, 2012; Alami *et al.*, 2021) also launched new development-oriented digital projects in other countries, to do good, but also in order to build their countries' diplomatic profiles elsewhere. Most international development aid has historically been a means of manipulating soft power, and the reconfiguration of the international aid context in the twenty-first century has done little to change this (Easterly, 2006; Koch, 2024). What has occurred over the last 15 or so years though is a considerable increase in the number of new donors, both in terms of countries and private foundations. China's role globally, not least through its Belt and Road initiative (Dunford, 2021), has been very significant, but interestingly the government has left its digital interventions especially in Asia and Africa, largely within the hands of its digital corporations, such as Huawei, ZTE, and China Telecom. Within the digital tech sector, major new actors have been Saudi Arabia, the UAE, and, to some extent, Qatar, who have hosted or been major sponsors for important international digital meetings, such as the IGF in Riyadh in 2024, and the ITU's WTDC in 2014, Plenipot in 2018, and WRC in 2023 all in Dubai, as well as the WISE events since 2009 in Doha. Both the UAE and Saudi Arabia are also frequent sponsors of ITU events in Geneva, such as the WSIS Annual Forums. These provided powerful symbolic opportunities for the governments of these states to profile their digital credentials 'for good', and through this increasingly to shape global policy. Moreover, these states have also funded digital interventions on the ground, often claiming to be delivering on the needs of the most marginalised, but at the same time seeking to establish their leaders as significant influencers in the post-hydrocarbon era ahead and raising their personal profiles. Once again, the state 'me', often personified by its autocratic rulers, has become a powerful player in the digital tech for development stadium.

Another, very different, example of the international agency 'me' syndrome that has specifically sought to do good, is the Principles for Digital Development initiative, stewarded by the US-based Digital Impact Alliance (DIAL), and developed in part from the UK's Digital Design Principles (see vignette by Proctor, p. 90). The Principles for Digital Development were originally created in 2014 to 'serve as a compass for those working to promote sustainable and inclusive development in today's complex digital landscape' (digitalprinciples.org, 2024; see also Waugaman, 2016; Vota, 2024), and were revised and improved in 2024. Their wide consultative process ensured that many existing good practices were included within the nine key principles:

1. understand the existing ecosystem;
2. share, reuse and improve;
3. design with people;
4. design for inclusion;
5. build for sustainability;
6. establish people-first data practices;
7. create open and transparent practices;
8. anticipate and mitigate harms;
9. use evidence to improve outcomes.

By March 2024, 315 organisations had endorsed the Principles. It is, though, impossible to tell exactly how many had actually used any, let alone all, of the principles in their practical work. DIAL (2025) has recently produced a useful report based on survey data from 34 organisations (mainly from NGOs and the private sector, all but one from Germany, Kenya, and the USA) across 18 implementation countries, as well as stakeholder interviews, that shows the effectiveness of the Principles in a small number of specific policy and practice contexts. A fundamental problem, though, with all such well-intentioned initiatives is that they also very much represent the interests of those who created them. The Principles are thus primarily a US-led initiative, with only three of the 14 founding organisations (eight of which were UN agencies headquartered at least in part in the USA) not being based in the USA. While 200 organisations from across the world were indeed involved in the original consultation, and 300 in the 2024 review, the Principles were conceived and developed within a mindset framed by interests within the USA, including USAID and the Bill and Melinda Gates Foundation. It is striking, for example, that Chinese, Indian, or Brazilian entities were not among the original founders. The top-down controlling ideology of such approaches, albeit well intentioned and producing good work, is also reflected in the way that 'the UN Innovation Network, for example, now requires UN entities to endorse the Principles as a condition for network membership' and 'The Principles have also served as a foundation for numerous other frameworks and standards, including the USAID Digital Investment Tool and UNDP Digital Standards' (DIAL, 2025).

The right people, building things they understand, and striving to deliver directly for citizens

Jamie Proctor

(Digital transformation specialist focused on the public sector)

In my first tech job, at a start-up in London, I was asked to review the processes for managing clients. Fresh-faced and excitable, I unwittingly made three mistakes:

- focusing on things that sounded exciting but didn't directly solve problems for clients;
- saying I had 'designed' something, when I hadn't comprehended the complexity of the problems I was trying to solve;
- trying to push through a 'vision', without the experience or expertise to back it up.

The CEO politely told me it was the wrong thing to do, and over the years I realised that he was right. I was lucky that I was working at a cash-strapped start-up, because had it been in the public sector, the answer might well have been very different. Public sector blind faith in buying big ideas has led to notable failures, such as the Post Office's Horizon initiative (Dinesson, 2024), or the £10 billion Future NHS failure (Baumann, 2021), both in the UK.

However, in the early 2010s, the UK built on success from the private and public sectors to develop the Government Digital Service (GDS), driven by a mantra of 'multidisciplinary teams, delivering iteratively, with a focus on citizens' (Lane Fox, 2012; Williams, 2012). The key change was to ensure that successful interactions with citizens (or users) should be the measure of success, rather than the amount of money spent on a product, or a manager's confidence in how things are going. The result was the best public sector website in the world (https://gov.uk).

There have been many attempts to transition the UK's experiences to low- and middle-income countries (LMICs), notably through the Principles for Digital Development (DIAL, 2025; see p. 89), which were based in part on the UK Service Design Principles (GDS and GDDO, 2012). However, creating multidisciplinary teams, delivering iteratively, with a focus on citizens in these contexts is even more challenging because:

- donor funding forces implementers to buy fixed outputs, thus preventing Agile software development;
- entrenched civil service hierarchies limit interactions with citizens, in preference for the latest management priority;

- it is difficult to recruit small multidisciplinary teams, despite expertise being available.

To serve the needs of the world's poorest and most marginalised, it is essential to systematise these changes towards *multidisciplinary teams, delivering iteratively, with a focus on citizens.* Otherwise, the potential benefits of digital technology in the public sector will never be realised.

The civil society 'me' and the challenge of 'projectitis'

Most civil society organisations have traditionally been seen by the public at large, almost by definition, as being altruistic and doing good, providing checks and balances between the powers of governments and private sector companies, and also frequently serving the needs of the marginalised, who have been failed by other kinds of organisation. However, this view has long been contested (Wood, 1990; Chambers and Kymlicka, 2002; Hamilton, 2003; NormCoresky, 2017; Brechenmacher and Carothers, 2018), and it must be recognised that civil society organisations (including NGOs) are not *necessarily* any better or worse than companies or governments. This is especially so in the digital tech for development context.

This section highlights three problematic areas, associated with the 'me syndrome' in civil society: (1) the individuals behind their creation; (2) the effects of short-termism associated with 'projectitis'; and (3) the agency of such entities in promoting private sector interests. This is not to deny the very positive role they often play in trying to support the poorest and most marginalised through the use of digital tech, and that many have been very successful in so doing (see vignette by Longmore, p. 27), but it is to try to understand why, despite their good intentions, civil society organisations working in the digital tech for development context still often do not achieve this. In so doing it recognises that many civil society organisations do indeed make some positive impact, especially for their beneficiaries and staff. The activities of organisations such as the Association for Progressive Communications (founded in 1990) and the Just Net Coalition (established in 2014), especially in its early days, have also been seminal in providing platforms for critical discourses around digital tech and development, and in trying to influence global policy that has been shaped mainly by the interests of the private sector and governments. They serve as a powerful reminder of the crucial role that civil society must play in global agreements on digital tech and development (see Chapter 7). However, all too often, civil society organisations, particularly those working in digital tech, still suffer from the 'me syndrome'.

The 'me' for 'we'

Most of the thousands of civil society organisations created to support poor and marginalised communities across the world through the use of digital tech have been created by people who are not members of these communities themselves. They are usually founded by those who are not very poor and can afford to be altruistic, although paradoxically evidence also shows that in practice the poor are often at least as altruistic as the rich (Dimick *et al.*, 2016). Much of the discussion above about the 'altruism' of those within large tech companies also applies to those who create civil society initiatives to try to use digital tech avowedly in the interests of those less privileged than themselves. Many motivations for such actions are usually interpreted as being positive (Mattis *et al.*, 2009), including religious beliefs, moral choices, political values, practical experience, and social conscience, but these are often also balanced by what tends to be seen as less positive reasons, such as a desire to have power over others, guilt, and sometimes even selfish ambition (Rachlin, 2002).

In practice, the creation of most digital tech-focused civil society organisations requires an individual or small groups of people to want to create them, and there is therefore always an element of selfishness in their creation. It remains at least partly about 'me', although focusing in intent on the needs of others. Getting this balance right is extremely difficult, typified, for example, in the context of global meetings and conferences, where civil society organisations are almost always represented by their leaders or senior staff, whose profile is thereby raised, rather than by the beneficiaries. This also applies particularly strongly in the academic context, where researchers working 'on' poor and marginalised communities are usually the ones to present papers at international conferences and thereby enhance their careers, rather than members of those communities themselves (see vignettes by Harindranath, p. 96, and by Juliette Unwin, p. 85; Unwin, H.J.T. *et al.*, 2023).

There are two important consequences of the 'me' in civil society organisations focused on digital tech for development. First, although motivation and funding are usually quite high in the early lives of such organisations, these both tend to decline over time as a result of challenges in implementation, underestimates of financial requirements (Flyvbjerg *et al.*, 2022; see also vignette by Gardezi, p. 34), the changing priorities of those involved, and the speed with which new, more advanced and deliberately disruptive technologies are developed to provide alternative solutions to the problems the organisation had originally set out to resolve. Successful organisations adapt and apply for new funding on different topics (see the section on projectitis, p. 93), but many do not survive and any benefits of the interventions upon which their beneficiaries relied are lost. It is extremely difficult to gain precise figures for the life-span of such digital tech for development organisations, but more generally Bond (the UK network for organisations working in international development) estimated in 2020 that only about half of its members would survive until 2022, albeit in part because of COVID-19 (Bond, 2020; see also Mettler and Miscione, 2022; Chaney and Jones, 2022). The dismantling of USAID by the Trump administration in 2025 and the reduction in funding from other traditional donors to enable them to reprioritise their budgets towards defence spending in response to increased levels of global warfare, especially in Ukraine, have dramatically curtailed the work of many digital tech-based organisations that it funded. This highlights the dangers for civil society organisations of relying on external, donor funding.

Second, one of the motivations of those establishing or working in such civil society organisations is in part their own personal fulfilment. Those who are successful frequently then receive employment offers elsewhere from larger organisations, donors, and even companies that are difficult to refuse, and their expertise is thus often lost to their prior organisation. Where such people are founders of a small civil society organisation, this can have a catastrophic impact on its subsequent ability to deliver effective interventions. Moreover, the interests of those involved in running a civil society organisation are usually not exactly the same as those of the poorest and most marginalised whom they intend to serve, and this can be the source of numerous problems in terms of delivery and outcomes (see vignette by Nwakanma, p. 16).

The challenges of projectitis

Projectitis (Gordon *et al.*, 1998) has been defined in various different ways, but two themes are at the heart of this usually fatal disease in the context of successful outcomes for digital tech for development initiatives: first, most such projects tend to be short-term, thus trying to ensure that lasting beneficial outcomes are achieved becomes extremely difficult; and, second, those running such projects often become so focused on delivering through the project management system they are using that they fail to take sufficiently into consideration the wider circumstances and ultimate intended outcomes.

One of the most critical challenges facing civil society-led digital tech for development projects on the ground is that they are usually only initially funded for three, or, more rarely, five years. Not only is this far too short a time to deliver any real beneficial outcomes in the most difficult and marginalised contexts, but also it means that the organisation is continually seeking new funding to maintain their staffing levels, and once a project is finished, employees often then move straight onto the next project without having the necessary time to see the original initiative through to fruition. These challenges are closely linked to the oft-cited recommendation that projects should concentrate on picking the low-hanging fruit, so that they can be assured of delivering success in reporting to their funders, thus helping to guarantee future funding for the organisation's next projects. All of these factors make it less likely that the poorest and most marginalised will benefit from such digital tech interventions, even if they are well designed and originally intended to benefit them.

Who benefits most?

All too often, civil society organisations rely on collaboration or formal partnerships with the private sector to implement digital tech interventions on the ground. Where these partnerships are well managed and overtly focused on delivering beneficial outcomes to the poorest and most marginalised, then a fair balance can sometimes be achieved, but these must be based on clear and transparent inputs and benefits frameworks (Unwin, 2005b; 2015a). Private sector companies can indeed provide valuable advice, experience, management expertise, and relevant digital technologies, and it is right to expect them to have a fair return on their investment. However, more usually, in the absence of clear ethical guidelines and transparent partnership agreements, these interventions serve the interest of the companies, who use civil society organisations as a conduit to gain access to previously untapped markets, to experiment with untried technologies, to acquire large amounts of data to enhance their business models, and to provide access to governments and policy-makers (see Note Three in Chapter 1).

This happens at all scales, from private sector sponsorship of large international conferences, to small-scale projects on the ground. The US-based Catholic Relief Services' annual ICT4D conference is typical of this process, with its main sponsors in 2024 being the Bill and Melinda Gates Foundation, Bixal (a US consultancy company), DataKind (a US-based AI and data science organisation), Esri (US producers of GIS software), Microsoft (US digital tech conglomerate), Nethope, RedRose (a technology service provider headquartered in London), and Sahaj (an AI, data and software company), with the US headquartered IEEE, USAID, and the Office of the UN Secretary-General's Envoy on Technology also being among the lower categories of sponsor. None of these, mainly US-based, companies would sponsor the event unless they saw real practical benefits for them in so doing. Although all such gatherings only flourish through the generous support of companies and governments, real questions must be asked about the ultimate practical benefits on the ground that they deliver for the poorest and most marginalised.

Many civil society organisations are also funded through the CSR (or CSER) budgets of major corporations which often enable them to provide valuable practical interventions for a specified period of time (see also Chapter 7). However, with all such project-based funding, it is extremely difficult to evaluate the relative balance of benefits between the companies and the ultimate intended beneficiaries. Furthermore, CSR financing usually does not provide structural change on the ground to enable an intervention to be sustainable in the longer term to the benefit of the poorest and most marginalised. For this, it is often better for civil society organisations to engage instead directly in realistic partnerships with the business end of such companies, rather than their CSR departments, since it is people in the profit-oriented sections who know best how to deliver continued sustainability and have the technical expertise to help deliver an intervention that might indeed become commercially viable, and thus sustainable (see vignette by Hughes, p. 115). The challenge here is that a well-intentioned digital intervention might also have adverse unanticipated outcomes. For example, giving mobile devices to poor people who have not previously used them as part of a development intervention in health, migration, or finance, not only turns them into data sources for governments and private sector companies, but always also carries the risk of introducing them to the many harms associated with their use, such as identity theft, sexual harassment and trafficking, or gambling.

'We' rather than 'me'

This chapter has argued that one of the main reasons why digital has not been created in ways that explicitly serve the interests of the poorest and most marginalised is that a majority of those in the private sector, and, to a lesser extent, also those in civil society, governments, and international organisation have their own interests at heart, more than they do those of the poorest and most marginalised. Furthermore, such initiatives are only very rarely crafted from their inception *with* the intended beneficiaries, but instead are developed by outsiders *for*, or worse still *on*, the poor and marginalised, with insufficient understanding of their real needs and the context within which they are to be implemented (see vignette by Harindranath, p. 96, as well as Chapter 5). As Seth (2022) has clearly articulated, change will only therefore come when enough people involved in the sector have a change of heart, and design digital tech interventions explicitly with, and in the interests of people in these communities (see vignettes by Nwakanma, p. 16 and Juliette Unwin, p. 85; Unwin, 2017a). It also requires a large dose of humility, which is often absent among those educated in elite institutions, who have always been praised for their success, and find criticism hard to take (see vignette by Aggarwal, p. 81). Many of these people really do think they know best, and as a result most continually fail to serve the world's poorest and most marginalised effectively (see epigraphs to this chapter). As noted earlier in this chapter, the Dunning-Kruger effect is rife in digital tech for development interventions with those leading them often being too inexperienced or ignorant to make wise decisions (Kruger and Dunning, 1999).

A small number of organisations are nevertheless both well intentioned and are trying to reduce inequalities through the use of digital tech, but there are far too few of them. There are far too many entities that have little or no real interest in designing and using digital technologies to serve the interests of the most marginalised in our societies; they focus on the next billion rather than the first billion (in my use of the term). It will be very difficult to change their minds in the short term. The majority of digital tech for development entities claim to be well intentioned, although they may in reality also be blind to the real, and unintended, impacts of their work. Many are more concerned with their own profile and profits at an individual and institutional level, or are simply unaware of the detrimental effects caused by their digital interventions. These are the people for whom this book is most intended; they are the ones who might change their actions if they really understand the outcomes that they have. Even by making a small shift, by thinking about the impact that their work has on the poorest and most marginalised and then trying to do something about it, can have a powerful effect. Many of the harms caused by the designers of digital tech could be reduced merely by explicitly considering in greater depth the possible unintended consequences of their innovations and acting to prevent them before they are created. Changing our language, so that we no longer use terms such as 'bridging the digital divide' or 'Internet penetration' can also be a powerful beginning (see Unwin, 2022a). Above all, we need to learn from the experiences, both positive and negative, of those who have worked before us on these issues, and place the 'we' before the 'me'. The biggest impact of all will come from focusing explicitly on equity and inclusivity of all digital tech 'solutions', rather than mainly on their impact on growth and our own personal and organisational interests.

How may academics help to empower marginalised communities through digital tech?

G. 'Hari' Harindranath

(Professor at Royal Holloway, researching societal impact of digital technologies)

An Indonesian migrant in Malaysia once asked a group of us during a research visit:

> Academic researchers and people from civil society organisations often come and ask us to answer surveys and questionnaires. But we never see them again when they go away, nothing happens and our lives remain the same. When is this going to change?

What, then, is the societal impact of much academic research, not just that on digital tech? How can academics ensure that their research addresses the concerns of the marginalised? These are questions we must grapple with when we make the decision to research 'with' marginalised communities, and not just 'on' them.

Ensuring positive outcomes for those we work with has many challenges, among the most significant of which are:

- *Prioritising outcomes over outputs*. Emphasising the priorities of those we work with will require a qualitatively different type of engagement than permitted by the existing dominant 'publish or perish' academic culture.
- *Commitment of time and resources*, including funding, beyond project timelines. Combining relevant research with teaching and other academic commitments is challenging for many academics.
- *Continued engagement with stakeholders*. Maintaining links with those with whom we work so that we can trace impact, ensure sustainability and learn lessons through monitoring and evaluation is easier said than done when most civil society organisations also depend on gaining new funding for their activities and indeed survival.
- *Evidencing impact and outcomes*. Impact can materialise in ways we never imagined at the outset. It is extremely difficult to measure the myriad routes through which our work shapes outcomes, not least through the cascading effects of conversations and new relationships forged.

In our recent research-practice on the safe, wise and secure use of digital technologies (https://ict4d.org.uk/sws), we worked 'with' migrants and other vulnerable groups in multiple contexts, prioritised contextually relevant ways of working, and focused on their needs (https://ict4d.org.uk/migration-and-digital-tech/). In Nepal, our emphasis was with returnee migrants, migrant and tech organisations, as well as community radio to reach isolated communities. In Brazil,

Figure 4.2 Sharing resources developed with the LBT community in Maré, Brazil

Source: Heloisa Melino, with permission.

we worked with LBT women and those living in favelas (Figure 4.2). In South Africa, we worked with migrant communities facing a hostile environment. In all of these instances, we tried hard to emphasise outcomes from the start, but this was challenging. In Nepal, we facilitated the development of a migrant information portal and training resources to be used at Migrant Resource Centres across the country. In South Africa, we trained groups of migrants in video-making skills and online safety that they are passing on to others in their communities. In Brazil, the LBT women we worked with created beautiful, localised resources for marginalised groups 'at the peripheries' in favelas, based on our initial training materials on online safety. Nevertheless, so much more needs to be done. We have continued to track the progress of these interventions, but gathering evidence of impact and sustaining them require considerable investment in both time and resources for both them and us as researchers.

Empowering marginalised groups through our work with digital technologies, and striving to make the world a better place as a result, may be lofty aspirations, but they are worth pursuing. That will require us all to get out of our comfort zones and find ways to prioritise outcomes, commit time and resources, and engage with communities on the ground, rather than in the ivory towers, to learn and gather evidence of impact and outcomes of using digital tech in the service of the world's poorest and most marginalised.

5 The innovation fetish

Tim Unwin

Innovation in digital technology is transforming international development by providing new solutions to global challenges. Through advancements like mobile connectivity, AI, and blockchain, technology is enabling greater access to education, healthcare, and financial services, especially in underserved regions. Digital platforms facilitate data-driven decision-making, improving the efficiency of aid distribution and enhancing transparency. Additionally, innovations in telemedicine, e-learning, and digital payment systems bridge the gap in remote or resource-limited areas, empowering communities and fostering economic growth. As technology continues to evolve, its role in advancing sustainable development goals becomes increasingly crucial in shaping a more equitable and connected world.

(ChatGPT response to request to draft 100 words on 'innovation in digital tech for international development', 8 April 2025)

The idea that digital tech innovation is, and should be, central to 'development' has emerged particularly strongly since the mid-2010s, hand in hand with the increased power of a few individuals, and the wider 'me' syndrome' discussed in Chapter 4. These innovations are invariably seen as being 'good', almost by definition. This is beautifully illustrated by the AI-generated text in the epigraph, which only mentions the positive potential of AI innovation. As with most existing AI systems, it draws on the dominant literature (data) that its algorithms and models have used to generate the text, and fails to incorporate any negative impacts of innovation in digital tech for development. However, such innovations do not always serve the interests of the poorest and most marginalised. Why did this text not mention the harms: AI for Bad, rather than AI for Good (Figure 5.1)? Moreover, it persists in promoting the myth that gaps and divides can be bridged. As the pace of innovation increases, so too does the risk of increased inequalities and the poorest becoming ever more marginalised. Digital tech innovations are also often positively described as being 'disruptive', but such disruptions are frequently harmful to those who can least afford them. It is often the poor and marginalised who are excluded from and exploited further through digital tech innovation.

DOI: 10.4324/9781003598008-5

Figure 5.1 Digital enhancement of posters at the AI for Good Summit, Geneva, 2025. The originals read from left to right, AI for Good, The Future is Here, and AI for Good.

Source: Tim Unwin.

This chapter challenges the often taken-for-granted assumption that innovation, especially in the field of digital tech and 'development', is *necessarily* positive for the world's poorest and most marginalised people. It begins by tracing the emergence of 'innovation' as a dominant leitmotif in international development, and then examines how a series of innovations in digital tech have successively replaced each other as the next best thing during the twenty-first century – without the poorest necessarily benefitting. It subsequently explores in detail some of the intricacies of digital innovation before re-emphasising that in most instances poor and marginalised people cannot afford failure. Agile approaches to international development by failing faster and breaking things may be less effective than efforts to extend what practitioners already know works well to those areas or people who do not already have them. Electricity, for example, is vital to any digital tech development, and yet worldwide 750 million people, most of whom are in Africa, did not have access to electricity in 2023 (IEA, 2024). A first simple step to ensuring that the poorest and most marginalised can benefit from digital tech is to make this basic infrastructure available and affordable for them. All this needs is the will to do it, rather than waiting for some new innovative digital tech, as the vignette by Banks (p. 100) illustrates so appropriately (see also Banks, 2013b).

***kiwanja.net*: Memories of innovation for the most marginalised**

Ken Banks

(Empowers global change through technology, innovation and storytelling)

kiwanja.net was born out of my belief that small, locally-led organisations, invested in the problem they are trying to solve, are often best placed to solve them. However, few in the 'official' development sector seemed to be giving them the chance. In the 1990s and 2000s, I became increasingly frustrated by how 'global development' often overlooked grassroots efforts. Experts were in the West, and everyone in the South was a 'recipient' of their shiny, donor-led innovations. Indigenous knowledge was often ignored, despite what ICT4D experts said on Twitter and at development conferences. By 2003, I'd already seen what mobile technologies could bring, and I decided to commit my energy to making sure that grassroots organisations were not left behind.

I was determined not to be distracted by shiny new things, and wasn't interested in what was or wasn't 'innovation' – I was just interested in things that worked. I believed strongly that we needed to strip away complexity and give people tools they could easily use and own on their own terms. That's what led to *FrontlineSMS*, one of the first truly offline, grassroots communication tools, enabling NGOs, activists and concerned citizens anywhere to connect and mobilise without needing the Internet (Figure 5.2).

Figure 5.2 FrontlineSMS in a school in Malawi

Source: Ken Banks.

I worked for 15 years trying to give a voice to, and support, the work of grassroots organisations through digital tech, but my frustration in a wider development system that didn't seem to want to do what they knew was best for those they were meant to serve eventually forced me to step away. Innovation was being driven more by donors than by the needs of communities, vast amounts of money were being spent hiring mobile and innovation experts, and I grew tired of the contrast between what ICT4D projects said they were doing, and what was happening on the ground – which wasn't much, in many cases. I no longer wanted to be associated with a sector that was badly letting down those it was there to serve. I still cared deeply about impact, about justice and equity – but I needed to find a new way to contribute. Stepping away wasn't easy, but it was the right thing to do.

How 'innovation' for growth is increasingly driving the world of development

Innovation is defined in a myriad of ways, often primarily to suit the specific discourse under consideration. Bound and Ramalingam (2016, p. 11), for example, suggest that it is 'the successful exploitation of new ideas that create value at scale', whereas Kumpf and Jhunjhunwala (2023, p. 15) suggest that it is 'synonymous with building institutional capabilities that enable an organisation, its staff and partners to leverage a specific approach or a technology when appropriate'. The latter of these highlights the emphasis that is usually placed on new technologies being at the heart of innovation, and this is the core focus for this section. The growing importance of both digital tech and innovation in development practice over the first two decades of the twenty-first century has been remarkable, reflecting in part the wider transformations in society caused by the design and use of new digital technologies. For example, the term 'innova-', as in innovate, innovation, and innovative did not appear at all in the Millennium Declaration in 2000 (UNGA, 2000), but it appears 61 times in the Summit of the Future outcome document of 2024 (UN, 2024b). This section shows how this rise in emphasis on innovation has occurred, and why the digital tech sector has played such a key role in it.

I first became aware of the sudden arrival of the 'innovation mob' (see also Bullinger *et al.*, 2009, who use the term in a slightly different way) when I was sitting in a meeting of the UN's High-Level Committee on Programmes (HLCP) in Rome in 2018, discussing issues around frontier technologies, AI capacity development, the future of work, and the future of education and learning. It all felt slightly rushed, because we had to leave time for a presentation about 'innovation', building on the establishment of the UN's new innovation network (UNIN) in 2015 (https://www.uninnovation.network). A very slick presentation followed, youthful, enthusiastic, and full of promise. The UN Secretariat had become persuaded that innovation, by which was largely meant innovation through digital tech and for economic growth, was the solution to the world's most pressing problems. Liu Zhenmin, Under-Secretary-General at the UN's Department of Economic and Social Affairs captured this well in his summary of the benefits in his Innovation

Tribute in 2019 on the Network's site: 'Innovation is a key tool of leverage for accelerated implementation of the Sustainable Development Goals' (Zhenmin, 2019). Trying to achieve the SDGs was the driver, and it is no coincidence that the Principles for Digital Development, discussed in Chapter 4 'form the shared values of all UNIN members and create a basis for joint innovation efforts. UNIN organisational members have adopted these Principles and committed to work in line with them whenever possible' (UNIN, 2023); it is chaired by the World Food Programme (WFP), the United Nations Development Programme (UNDP), and the United Nations Children's Fund (UNICEF). Little evidence is available, though, on how successful such practice has been. Some agencies within the UN system had long championed innovation for well over a decade before the creation of the Network, and it was designed in part to unite these disparate voices and better serve the interests of those promoting innovation. UNICEF's Office of Innovation was thus first established in 2007, and it has commendably sought to learn from failures as well as to champion successes (UNICEF, 2014; Ramalingam and Bound, 2016). UNCTAD has likewise published its influential *Technology and Innovation* Report series since 2010 as a spin-off from its earlier *Information Economy Report*, although its interests in technology transfer can be traced much further back to the 1960s, and included an element of innovation from the 1980s onwards (UN, 2004). UNESCO was likewise active with its Global Observatory of Science, Technology and Innovation Policy Instruments in the mid-2010s (see also Moreno *et al.*, 2024).

It was not, though, within the UN system that 'innovation' as a key theme first appeared within the field of international development. There has been a long history of intellectual and practical interest in innovation diffusion, not least through the foundational work of the geographer Hägerstrand (1953) and the sociologist Rogers (1962), as well as the controversies surrounding India's Green Revolution in the 1960s and 1970s. This significantly improved grain productivity but was accompanied by increasing inequalities and environmental damage (Bayliss-Smith and Wanmali, 2009). This research and practice showed clearly that unless mitigating actions are taken, innovations have a tendency to increase inequalities between those who can afford them and the poor who cannot. To use Roger's terms, innovators and early adopters are likely to have a competitive advantage over the laggards (although for a critique, see Unwin, 1987).

The USA was at the forefront of much of the early enthusiasm around digital tech innovation and development, building on lessons learnt and funding increasingly available after the dot-com bubble of the late 1990s and fears over the Y2K 'problem' (Manion and Evan, 2000). It was also in the Washington-based World Bank that some of the clearest evidence for a new digital tech-related emphasis on innovation was most evident, especially in the context of the deregulation of the telecommunication sector. The World Bank's ICT sector strategy thus comments that 'ICT sector reform involves changing the overall policy, legal, institutional, and regulatory framework with the objective of opening markets to private investment and competition, thus fostering *innovation*, efficiency, and improved services' (World Bank, 2002, p. 3; my emphasis). Some of this was to be implemented through the *info*Dev programme, a partnership of donor agencies, companies, and civil society founded in 1995, with its secretariat in the World Bank (DFID, 2019), which subsequently grew into one of the most important sources of rigorous evidence on ICTs and development in the 2000s. One of its main programmes was the Incubator Initiative, launched in 2002 with the support of the government of Japan (*info*Dev, 2005). This was 'dedicated to establishing a network of incubators to facilitate the emergence and development of small and medium size enterprises (SMEs) in developing countries that utilize ICT to improve performance and expand operations' (*info*Dev, 2005, p. 7). At the heart of this programme was the fostering of ICT-enabled innovation and development in developing countries; by 2005, it already had a portfolio of 43 projects across the world.

At the same time, USAID launched its first programmes with US universities to serve global horticultural needs in 2009, which later evolved into the Feed the Future Innovation Lab for Horticulture (UC Davis, 2023). More recently, USAID's Development Innovation Ventures was founded in 2011 as an 'open innovation program that tests and scales creative solutions to any global development challenge' and in a period of just over ten years had given $260 million to 327 awards (USAID, 2024). Likewise, in the UK, the government created a Global Innovation Fund in 2013, 'To identify and support innovative solutions to development challenges' (FCDO, n.d.). The seven UK research councils were also merged with Innovate UK (originally founded in 2007) into the new body UK Research and Innovation in 2018 (UKRI, 2019), and post-COVID-19 the UK created a new innovation strategy, stating quite clearly that, 'We are in a race to the top. The UK government, therefore, has an overarching goal of making the UK a global hub for innovation, placing innovation at the centre of everything this nation does' (DSIT and DBEIS, 2023). Again, the 'me' syndrome raises its ugly head.

This overt focus on innovation to enhance economic growth that emerged in the international development sector between the late 2000s and 2020 has been of two broad types: (1) innovation in the delivery of 'development' using digital tech to replace earlier types of intervention, as with early efforts to replace books by innovative digital devices, such as OLPC (see Chapter 4, and the vignette by Hollow, p. 126); and (2) innovation through the encouragement of digital tech within economically poorer countries themselves, as with incubators and innovation hubs intended to enhance economic growth, but also tying these countries yet further into the economic systems dominated by global capital located in the USA and Europe, and more recently China.

There are numerous, often context-specific, reasons why this emphasis on innovative digital tech for development emerged so strongly in the second decade of the twenty-first century (see vignette by Scur, p. 104). Prime among these are the motivations of the global technology companies situated mainly in the USA, but also in China, and their growing influence over UN agendas, as discussed in Chapter 3. These were also underlain by the 'me' syndrome and the overconfidence of the leaders of these global corporations, large civil society organisations and international agencies discussed in Chapter 4. All too often, civil society organisations and UN agencies did not have sufficient expertise in or knowledge about these technologies and were thus easily beguiled and led astray by the corporations that promoted them. However, there were also other factors that came into play. By the 2010s, many new young aspirational graduate employees who had direct experience of these technologies in their daily lives, who had been educated with, through, and about them, and who still retained a degree of altruism, were beginning to be employed by donor agencies, civil society organisations, the major consultancy firms as well as within UN agencies. For them, using digital tech in their work, having seen the perceived benefits that it brought for them, and wishing to share these more widely, it seemed obvious that there might be significant benefits in using digital tech in the 'development' sector (see vignette on Imfundo, p. 11 in Chapter 1, for an early example). This focus on using innovative technologies would have happened, regardless of the efforts by the tech companies to promote their sales into new markets, but at a somewhat slower rate. For the few who had time to read it, this digital innovation focus was also promoted by the findings of increasing amounts of research, suggesting that the funding of innovation within the economically poorer countries of the world could also bring significant economic benefits. Kremer *et al.* (2021), for example, suggest that in the early stages of USAID's DIV (Development Innovation Ventures) programme, each dollar spent generated at least $17 in social benefits, with those reaching the most users most likely to have low unit costs, to leverage existing initiatives and to be grounded in research. This in turn further encouraged organisations and governments across the world to invest more in innovation hubs. However, most such arguments were framed largely in economic terms, and failed sufficiently to examine their social implications in terms of equity. A further factor that encouraged donors and international agencies specifically to address digital innovation was the rapidly increasing demand from their member governments for guidance and help in all matters relating to the role of digital tech in their economies. 'Enlightened' governments in Africa, Asia, and South America sought help and advice about innovation and regulation, although there were also numerous instances where senior government officials often felt it was easier simply to accept the offers being made by the major tech companies keen to invest in their countries (*The New Times*, 2008).

Digital tech and the unbanked: The COVID pandemic in Brazil

Fernanda Scur

(Systems thinking for active collaboration, partnerships and collective impact strategist)

Digital tech Initiatives often claim to use cutting-edge technology as a panacea to solve problems, without considering that the problem they are trying to solve is a wicked and complex one. Any digital solution must also therefore be complex and, above all, take into consideration both the context for which it is intended and the stakeholders that need to be involved in its implementation.

Brazil's federal government offered financial emergency aid to informal workers and small entrepreneurs during the COVID pandemic in the early 2020s (Monteiro et al., 2020; Arruda *et al.*, 2021; Sordi, 2023).

This could be requested quickly through a banking app or more slowly through a visit in person to a bank. The numbers who went to banks revealed that 46 million Brazilians were previously *invisible* to the government, mainly because of a lack of Internet access and digital skills, but also the large numbers of people who did not have an official digital ID (CPF) and were therefore unbanked. The poorest, who needed the aid the most, were often the ones who had to wait the longest, and frequently never received the financial support.

This issue surfaced almost by chance and demanded a quick turnaround response from the Brazilian government, which swiftly decided to partner with many other stakeholders from government, the private sector and civil society. What followed was an increase in the use of banks and financial instruments by the poorest. This demonstrates that an ability to analyse a context strategically, to make connections among the many possibilities available, and to include multiple stakeholders are all crucial in dealing with the uncertainty of an emergent issue, as in the example above. While the government has the scalability, it needs to collaborate with other actors that have the resources and knowledge that it does not have. Often it is the small organisations that have the most difficulty in going to scale, yet their role on the ground is frequently essential for the successful delivery of multi-sector interventions.

However, this example also has a darker side. The use of digital technologies ultimately forced many previously unbanked people to enter the digital world of banking and the wider digital economy, but it also increased the ability of the government and financial sector to surveil them much more easily and effectively.

The next big thing: ICT4D, incubators, EdTech, M4D, Blockchain, AI for Good . . .

The speed of technological innovation has been increasing at an ever-faster rate in the twenty-first century. In part, this is simply a result of the increasingly rapid demographic growth that has occurred over the last half-century, leading to there being many more scientists doing 'innovative' research in university and corporate labs. It is also a result of the dramatically increased speed of digital computation described by Moore's Law (the approximate doubling every two years of the number of transistors in an integrated circuit), so that digital tech itself is being used to speed up cycles of innovation. This transformation has for some been seen as a Fourth Industrial Revolution (Schwab, 2017; also known by some as Industry 4.0) on the grounds that it is affecting all disciplines, and is different in scale, scope, and complexity to anything that has gone before. However, it is still being driven by the same fundamental drivers of capitalism that have existed since at least the seventeenth century, it is grounded in technological determinism and instrumentalism, it represents a largely male heroic and elite view of history, and it is a self-fulfilling prophesy (Unwin, 2019; Moll, 2021). Some prefer the term 'Frontier Technologies' that began to be used increasingly in the late 2010s to refer to new technologies at the ever-moving frontier of knowledge (Spar, 2015; Ramalingam *et al.*, 2016; UNDESA, 2018). However, even this term remains problematic in the context of 'development', because it serves as a reminder of the 'American Frontier' in which white, initially mainly male, settlers 'heroically' conquered the Native Americans or First Citizens through the use of new technologies (Figure 5.3). This was anything but a positive frontier for the people who were conquered through use of new technologies (the telegraph, the rifle, and the railway). It must always be remembered that a marginalised person's new 'frontier', such as an old Android 'phone, is a backwater to the rich elite's AI-running smartphone frontier technologies.

There is a close analogy here between Gast's image from the 1870s and US and Chinese corporations currently driving the frontier of digital tech forward into Africa, Asia, and Latin America, exploiting the existing civilisations to be found in these places for profit (see Figure 5.3). Accordingly, every new wave of digital tech has become associated with a new and better technology-based development agenda: in the late-1990s, Fuchs at IDRC coined the term ICT4D (Elder *et al.*, 2013); in the early 2000s, incubators became increasingly popular; then, in the early 2010s, mobiles for development (M4D) grabbed the attention of development practitioners (Banks, 2013b); e-Learning became EdTech, placing the emphasis on the technology rather than the learning; and more recently we have had Blockchain

Figure 5.3 John Gast, American Progress, 1872. Chromolithograph, published by George A. Croffutt

Source: Prints and Photographs Division, Library of Congress, LC-DIG-ppmsca–09855. Available at: https://lccn.loc.gov/97507547.

for Development (Cunha *et al.*, 2021), and the latest 'big thing' taking the headlines being AI for Good (ITU, 2024) (see Figure 5.1). Cyborgs will be next.

Anyone attending the overlapping WSIS Annual Forum and the AI for Good Summits in Geneva in 2024 and 2025 will have noticed the differences between them. The AI for Good summits were 'new', they were glitzy, they had high profile 'me' personalities, they had loud music and bright lights, they were commercial, with private sector companies touting for business in the exhibition areas (see Figures 6.2a and 6.2b). Nevertheless, they had rather few people there who really knew much about the inequalities driven by digital tech in the poorer countries of the world. Such people in 2024 were left mainly in the 'dungeons' of the old ITU buildings across the road for the coinciding days of the summits, and in 2025 were in the much more low-key, table-top environments of an adjacent hall in Palexpo (Figure 5.4). As wave upon wave of 'new' innovative digital technologies are developed, the companies creating and selling them are eager to establish new markets, and they see the 'development industry' as a perfect means to generate greater profits (D4ICT). Moreover, given the criticisms that AI models have to date largely been built on biased data with respect to cultural and ethnic characteristics (O'Neill, 2016; Chen, 2023), the opportunity for these companies to acquire new large data sets from users in middle-income and less-developed countries is an important bonus. Furthermore, naming such summits as being 'for good', rather than 'for bad' (see Figure 5.1), helps to establish a significant new symbolic iconography in the minds of participants. As earlier chapters in this book have argued, no technology is inherently good or bad; drones can be used to rescue people or to

Figure 5.4 WSIS exhibitors in Geneva, 2025

Source: Tim Unwin.

kill them. It is the people who design, create, and use them who do good or bad; most people want to be seen as doing good, but invariably their innovations can do as much harm as they do good (O'Neill, 2016; Marcus and Davis, 2019; Floridi, 2023; Coeckelbergh, 2024).

If current practices are to be changed, and new digital innovations developed that take into account an understanding of past experiences in the interests of the poorest and most marginalised, then it is crucial to understand why each wave of digital innovation is so widely seen in a positive light. In part, this is because new technologies frequently offer novel and significantly improved benefits for users, albeit usually at a financial cost that further marginalises the poorest who are unable to afford them. This lies at the heart of the harmful and environmentally unsustainable business models of many digital tech companies. Equally importantly, though, there has long been a deep fascination with novelty and the idea that anything new is better, often referred to as neophilia (the love of the new), which remains powerful, even when it does not always provide real benefits. Many new things can indeed be better, but they usually carry higher risks that poorer people are not willing to take. It is not that these people are ignorant laggards, but rather that they are making wise decisions based on their circumstances and knowledge. This has interesting similarities with Chayanov's notion of how the peasant economy functioned in the nineteenth century (Thorner *et al.*, 1966).

More generally, Cloninger (1994) included novelty seeking as one of the four temperaments in his Temperament and Character Inventory, and it seems likely that dopamine activity in part of the midbrain is higher in people who are also high in novelty seeking (Zald *et al.*, 2008). This might suggest that the rewards from seeking out new experiences could also be related to addictive behaviour, thus reinforcing a possible link with addiction to new digital products among consumers. Neophilia can be both a transient state and an abiding individual characteristic (Gallagher, 2011), so while all of us can be excited by the new, the way we react appears to be conditioned by our genes, brain structure, neurochemistry, and life experiences. Whitbourne (2012) further suggests that novelty-seeking tends to be positively associated with exploratory excitability, impulsiveness, extravagance, and disorderliness, and it would be interesting to explore further the extent to which these characteristics underlie the behaviours of those championing the 'next big thing' in digital tech. One way to change the innovation fetish might actually be to ensure that more people who do not share these characteristics are involved in teams seeking to design and use digital tech in the interests of the poorest and most marginalised.

The fundamental problem with the creation of these waves of hype over the next best digital tech 'for development' is that they are instrumental and reinforce a belief in digital determinism (see Chapter 2). Most of these 'innovations' have been commercially driven, and are imbued with an underlying profit-seeking motive. They are not primarily designed with, or in the interests of, the world's poorest and most marginalised people, although companies fully understand that the 'development market' can certainly add to their sales and ultimately profits. Moreover, lessons from the previous cycles of the 'next best thing' are insufficiently learnt because they are associated with the old, rather than the new. This was well represented by the hype over 'Blockchain for Development' during the late 2010s and early 2020s (Cunha *et al.*, 2021). Zwitter and Boisse-Despiaux (2018, p. 1) thus commented that 'Blockchain technology is rapidly reshaping the digital tools we use to conduct daily transactions' and that it 'is swiftly entering the fields of humanitarian and development aid'. This is a classic example of a problematic instrumentalist perspective on technology, because it is not blockchain itself that is doing this, but rather the people who design, promote, and sell it who are shaping the tools. Many of the strongest proponents of blockchain for development had made fortunes in the early boom years of its use in cryptocurrencies, and were seeking new ways to use the technology and profit further from it; it was a solution in search of a problem. Significantly, the richest crypto and bitcoin billionaires are once again almost all men, and are mainly based in the USA (Forbes, 2024b). Many of them have worked hard to promote blockchain through their networks, and within the humanitarian and development aid sector. However, as Zwitter and Boisse-Despiaux (2018) again note, while the use of blockchain can have positive outcomes for development, it also has the potential to perpetuate social problems and to add new risks for those involved in development aid. This is true of every new wave of digital tech. As an accelerator, digital tech increases inequalities and is likely to harm at least some people unless mitigating actions are taken. Where inequalities are greatest, the use of digital tech is likely to increase harms the most for the poor and marginalised.

Not all agree with such a pessimistic perspective towards the newest technologies. Arora (2024) is, for example, much more optimistic, taking hope from the experiences of users in economically poorer parts of the world such as India, Bangladesh, and Brazil, where she notes that young people see digital tech as providing them with a rare space for self-actualisation. She thus comments that 'Anyone who has ventured outside the West will feel the contagious optimism from young people's extraordinary hunger for being digital' (Arora, 2024, p. 2). For her, the 'next billion users are the new BoP in the digital economy' (Arora, 2024, p. 10; see also Prahalad and Hart, 2002; Prahalad, 2004; Arora, 2019). In particular, she draws on AI for Good projects to emphasise the need to pivot from altruism towards relationally driven approaches to design that help explain 'how relations between people, places, and technologies can work in ways that sustain human dignity, civility and responsibility'. She is optimistic that despite the problems with current algorithms, this will change as young 'users from the majority world infuse these algorithms with their aspirations through their everyday digital acts' (Arora, 2024, p. 6). However, the digital neophilia by the young across the world who do have access to these technologies is partly driven by the interests of those designing the technologies in the first place, thus constraining the real autonomy of the youth of Africa and South Asia. Moreover, what may be true of the 'next' billion (Arora, 2019) is not necessarily true for the poorest and most marginalised, those that I term the 'first billion' because, for me, they are of most importance. Relatively wealthy young people across the world are swept up in the enthusiasm for technology, but the poorest are left behind, as reflected poignantly in Alok's cartoon (Figure 5.5).

Figure 5.5 Alok's cartoon about the inequalities associated with online education
Source: with permission from Alok.

It is also pertinent to recall that the lifeblood of the global AI industry is in the labour of very large numbers of people, especially the young, often working in conditions that have been described as modern-day slavery in countries such as Kenya, India, Nepal, the Philippines, and Venezuela doing the essential task of data labelling/annotation (Chandhiramowuli *et al.*, 2023; Pogrebna, 2024; Privacy International, 2024). While such labour exploitation is often rightly criticised, some companies have a better reputation than others, and it is particularly interesting to see the high number of data labellers from Nepal now advertising their services on LinkedIn and thus seeing it as a positive lifestyle in the mid-2020s.

How much digital innovation failure is acceptable in international development?

Failure is an integral part of innovation, but at the same time innovation is a central pillar of corporate survival. A report on corporate innovation by Kaspersky (2020, p. 14) thus notes that '95% of innovations fail before launch', with more than a third of these failing during the development stage. There are nevertheless very different views about the key factors that can help improve the success of innovations. Some prefer teams (Kaspersky, 2020), whereas others tend to prioritise the role of key individuals identifying things that others have missed (Krippendorff, 2019). Worryingly, there is also increasing evidence that scientific research is becoming less innovative and disruptive. Although there is exponential growth in the volume of scientific publications, Park *et al.* (2023, p. 138) suggest that 'papers and patents are increasingly less likely to break with the past in ways that push science and technology in new directions', mainly because there has been a narrowing in the use of previous knowledge (see also Chapter 4).

These observations raise a fundamental question: if most innovations fail, and real scientific innovation is declining, should we be focusing so much on digital tech innovation specifically for international development? If we do not really know what the impacts of innovations in blockchain, AI, robotics, and biomedical research are for the world's poorest and most marginalised, is it morally acceptable for us to be experimenting on these groups of people? The world's poorest and most marginalised cannot afford failure. This section explores three aspects of this dilemma: (1) the balance between success and failure in digital tech innovation; (2) the role of agile practices; and (3) learning from our mistakes.

Those who benefit from digital tech innovation, and those who are left further behind

This sub-section addresses two related but distinct issues: the location and intended beneficiaries of digital tech innovation; and the extent to which innovation-focused policies and practices have been successful for their stakeholders and intended beneficiaries. There are many examples of digitally innovative approaches to development that have had some beneficial impacts, especially when focused on economic growth, and much can be learnt from these (see vignettes in this book, especially by Hughes, p. 115). The WSIS Prizes, initiated in 2012, and voted for annually online by millions of participants in the WSIS process, provide a good perspective on the fields across all the WSIS Action Lines from which the 18 winners and 72 runners-up each year have been drawn. Categories that have historically attracted most attention include ICT Infrastructure, Capacity Building, Access to Information, e-Government, and e-Health. Equality, inequality, or equity do not feature prominently, in part because there was never a specific category for them in the WSIS founding principles agreed in Geneva in 2003 (WSIS, 2003, Section B). More broadly, areas where particular successes have been identified in the use of innovative digital tech for development have been achieved include the provision of government services (Araujo *et al.*, 2021; UNDESA, 2024), wider participation of women in the digital economy (see the EQUALS Awards), and the rapidly expanding 'EdTech' sector (see EdTech Hub, https://edtechhub.org/), although much further work needs to be done in all these areas with respect to how the most marginalised can really benefit (Unwin *et al.*, 2020). Even the most optimistic practitioners will nevertheless probably agree that the application of digital innovations in international development has been less successful than they would like to admit in public, especially in terms of systemically reducing inequalities and benefitting the poorest and most marginalised.

The distribution of innovative digital tech for development interventions globally and within countries is usually markedly skewed to the richer and more affluent areas and people (see vignette by Hughes, p. 115). Remote rural areas, ethnic minorities, those with disabilities, refugees, and women and girls in patriarchal societies (Buskens and Webb, 2014) all still remain marginalised in terms of the benefits that they receive from innovations in digital technologies (see vignettes by Okello, p. 51, and Tarmizi, p. 175). In part, this reflects the low-hanging fruit tendency already discussed in Chapter 4, but even the terminology often used can be seen to accentuate these differences. The notion of 'Smart Cities', for example, implicitly suggests disparagingly that villages are 'dumb'. Although there have indeed been some attempts to develop innovative community-led smart villages based on digital tech in richer parts of the world (European Network for Rural Development, 2021), those in Africa and Asia have at best had a mixed record of success (Pun, 2023; Smart-Villages, 2023; see also Chapter 4). They have also tended to last for a relative short time until the enthusiasm of their founders dissipates, and many have focused in practice more on wider development issues such as electricity, agriculture, and environmental impact (Smart Villages Research Group, 2017), than specifically on the use of digital innovations such as blockchain or AI. In contrast, there is a plethora of Smart City initiatives based upon the use of a range of innovative digital tech across Africa (Smart Africa, 2017; Warner and Ajibade, 2024), Asia (Yoon *et al.*, 2020; ASEAN, 2025) and South America (Alarcon and Verheije, 2024), attracting very considerable levels of investment and technological inputs. This can only but increase digital inequalities between those living in innovative smart cities and those in more isolated, poorly connected, and presumably 'stupid' or 'unsmart' areas.

Likewise, innovation in digital tech can transform the lives of people with greater disabilities far more than it can those with fewer disabilities (see vignette by Gardezi, p. 34). Yet despite this potential, it is the less-disabled who have in practice benefitted most from innovation in digital tech, especially in the poorest and most marginalised contexts. Despite well-established groups campaigning for inclusive digital tech, insufficient of them focus enough on the needs of people with disabilities and the injustices that mean that most people with disabilities still have to pay considerably more for the life-changing innovative assistive technologies that they need to survive in an increasingly digital world (Eid, 2017; Mills, 2024). Indeed, many broader 'digital inclusion' initiatives pay scant attention to the needs of people with disabilities. The Global Digital Inclusion Partnership, for example, tends to focus especially on the inclusion of women and girls, rather than on accessibility and disability (a search on their website in March 2025 only listed 8 results for disabilit-, in contrast to 31 for girls and 69 for women). This further accentuates the inequalities, even between communities that are already marginalised.

Doubts can also be cast on the real extent to which ambitious innovation-focused digital policies and practices, such as innovation hubs and Smart Cities, have actually been successful, even for their intended stakeholders and beneficiaries in Africa, Asia, and South America. Evidence varies greatly about the success of such initiatives, not least because of their differing purposes and criteria for success, as well as a failure to embed rigorous monitoring and evaluation processes within them. Innovation hubs have generally focused on a mix of providing a physical co-working place with appropriate infrastructure (hackerspaces), creating support networks for young entrepreneurs, providing training in both entrepreneurship and technical skills, and convening opportunities for young entrepreneurs to pitch to potential investors. Some individuals in some such hubs have undoubtedly been successful, and gone on to create effective companies, or have had their start-ups bought out by larger, often US-owned, corporations. However, as Pompa (2013, p. 10) notes, the literature suggests that, 'Overall there is little evidence of long-term sustainable impact across sectors with results of few credible studies suggesting that support to start-ups in a developing country context as a whole, have limited effectiveness' (see also DFID, 2019). Key challenges for incubator initiatives, especially in Africa, have proven to be their sustainability beyond short-term project funding, over-aspirational expectations, discontinuity of government support (especially when governments change), overemphasis on digital skills with insufficient provision of entrepreneurship training, and the lack of a wider economic base within a country to sustain potentially successful new businesses (Korsheed *et al.*, 2014; Ganamotse *et al.*, 2017). Despite these problems, the incubator movement has proliferated to such an extent that XYZLab (2025) lists more than 1,600 start-up accelerators and incubators across more than 100 countries. Among the best-known African innovation hubs are the iHub, launched in Kenya in 2010, which in 2019 became part of the Nigerian Co-creation Hub (CcHub), originally founded in Lagos in 2011. Egbetokun (2023, p. 16) suggests that such incubators remain an important tool for 'supporting the survival and growth of viable enterprises' and that 'Africa needs business incubators in its pursuit to reduce the burden of unemployment and foster industrial development' (see also vignette by Ileleji, p. 30).

Likewise, across South Asia, there are those who continue to advocate for the value of incubators and innovation funds. In Pakistan, the government inaugurated the National Incubation Centre in 2016 (Shafqat, 2016) and its National Technology Fund, Ignite, has subsequently promoted incubators to nurture start-ups focusing on the 'Fourth Industrial Wave Technology', based on the premise that 'innovation is the prime determinant of the rise and fall of societies' (Ignite, 2024). In India, the Science Technology and Innovation portal likewise listed 248 incubators in March 2025, and particularly encourages close collaboration between incubators and universities (Ministry of Human Resources Development, Government of India, 2019; ISTI, 2024). Governments undoubtedly need to provide support for training and skills development in digital technologies and entrepreneurship, but without rigorous global cost-benefit analyses it is extremely difficult to claim 'success' for such incubators and hubs. All too often, foreign companies support them in their own interests to identify future employees, to extract innovative ideas, and to be able to exploit cheaper outsourced labour (Holden and Harsh, 2024). It looks impressive that Nairobi's iHub (2024) can claim that 450+ start-ups trace their roots to the hub, that 5K+ individuals have worked with them, and that 1M+ people have attended their events, and that the CcHub (2025) has supported 90+ ideas and companies since its origins, with over 50 advisors, mentors, and facilitators a year contributing to its programmes. However, it remains unclear about what the longer-term outcomes of such hubs really are for the majority of their users, and especially what their outcomes have been for the poorest and most marginalised people in these countries.

Similar arguments apply to the contrasts between the rhetoric and reality of smart technology-led cities in many of the economically poorer countries of the world, typified in Africa, for example, by Konza in Kenya, and Moamba in Mozambique. The Konza Technopolis was conceived in 2008 as the country's 2030 Flagship Project (Konza, 2024) and launched in 2013 as a $14.5 billion project to transform Kenya into Africa's leading hub in the global software industry. A prescient paper shortly afterwards by Saraswati (2014, p. S218) commented that, 'Rather than facilitating the creation of a dynamic software service industry', it was likely to 'retard wider technical progress in the country'. Progress on the city has been painfully slow (Baraka, 2021), and by 2025 although the road infrastructure and layout of plots had been largely completed, there were only a handful of buildings there, most of which were still under construction (Sikolia, 2025). Despite this, there remain numerous eulogistic proclamations that it will soon be finished, helping to turn Kenya into a successful digital middle-income economy, as planned, by 2030. Moamba in Mozambique is even less successful than Konza. It was first mooted as a science park in the early 2000s and was one of the four such parks announced by the government in 2008 'to be hubs of forward thinking and innovation where science and industry come together to find appropriate and workable solutions to the country's problems' (Erasmus, 2008). This expanded in concept over the next decade, funded in part by Chinese and South African investment for a 900 ha Parque Agro-industrial de Moamba (Reidy, 2019), with discussions of turning this into a smart city in the early 2020s. Much of its anticipated success was tied to the construction of a dam on the Incomati River, first mooted in the early 2010s with funding from Brazil which would provide reliable water for agriculture. However, as of 2025, the dam's financing had still not been finalised, and there was still little to see of the new city on the ground.

Most such smart city initiatives, especially in the world's economically poorest countries, have failed to live up to the rhetoric of those promoting them. Even if they were never intended explicitly to serve the poorest and most marginalised people in these countries, 'smart' initiatives have mostly failed even to serve their richer, tech-focused, urban elites. Baraka (2021) rightly comments that part of 'The answer has to do with global management consultancies, such as McKinsey, and their relationships with the big technology firms of the West.' The McKinsey Global Institute (2018) has thus been at the forefront of selling the dream of smart cities, clearly exemplifying the link between innovation and the corporate 'me' syndrome discussed in Chapter 4. Likewise, the big tech companies of the USA and China are almost by definition also the biggest winners from the creation of anything claiming to be technologically 'smart'. This focus on 'me' is also evident in the actions of local and national politicians who have sought kudos among their electorates by creating visions of a smart urban future, as well as the landowners who benefit from the sale of their land or dramatically increased rental values. Having the support of the World Bank (2024b) through its Global Smart City Partnership Program, established in 2018, adds a considerable level of status and credibility to such schemes.

The power of micro-transactions

Nick Hughes

(How digital money can unlock powerful new business models)

Few can doubt that digital technology is a powerful enabler when used well. However, an obsession with digital scale for scale's sake is a mistake, driven in part by the venture capitalist culture of 'get big and profit will come'. In emerging economies, the 'scale population' simply does not have the disposable income from which to draw profit. Rather, we must use technology to create new economic wealth by either helping someone make money or save money: *market-creating innovation* (Christensen *et al.*, 2019). Critically, we also still need humans in value chains to mitigate the friction of low digital literacy. Boots-on-the-ground are still required (Wheeler Fintech and Scale Team, 2021)! These brief comments draw on my experiences in leading teams in M-PESA and M-KOPA that have designed, deployed and scaled relevant digital propositions.

There are currently over 2 billion registered mobile wallets worldwide, excluding crypto accounts (GSMA, 2025). We didn't have this term in 2007, but M-PESA is now a stable coin – an electronic creation of digital currency tethered hard to fiat money in a bank. With our launch slogan of *'send money home'*, M-PESA set the model for distributed financial services and has been shown positively to impact those below the poverty limit, letting them earn micro-income in the informal economy or providing a means to get cash quickly from friends and relatives in an emergency situation.

Sending a phone-based micro transaction quickly, safely, and from anywhere to anywhere at relatively low cost drives liquidity in a world where, even today, the majority of people are unbanked or underbanked. The next phase of development will see digital payments linked to real-time economic output data. The use-cases are multiple, with the following being just a few: clean energy from distributed solar; earnings from tiny retail outlets unlocking working capital; and the lock-up of carbon into biochar rewarding buyers and sellers in the carbon markets. These all have economic activity that we can capture (irrefutably) in data and therefore link directly to a digital payment (reward or incentive models work).

We use the term Activity-Based-Finance. This means *activity data linked to e-money*: the data provide proof-of-work that economic value has been created at a micro scale; and micro-payments reward that effort. Exciting new business opportunities lie ahead. Perhaps even an alternative to expensive interest-bearing credit? Surely this is a better, more equitable way to drive economic growth?

Agile approaches to digital tech and development

Innovation, particularly in the digital tech sector, is often associated with 'Agile' ways of thinking and project management. Advocates of Agile thus see it as being particularly valuable in contexts where there is uncertainty, and the precise problem being addressed is not necessarily initially known. As noted at the beginning of Chapter 4, Zuckerberg is famous in the early days of Facebook for promoting the company's motto as 'Move fast and break things . . . Unless you are breaking stuff . . . you are not moving fast enough' (*Business Insider*, 2009). This reflected his views that to be successful it was important to be agile, nimble, to take risks and to think outside the box. However, by 2014, he and the company had matured and the internal motto changed to 'Move fast with stable infrastructure' (Baer, 2014). While Facebook may have repositioned itself (Wynn-Williams, 2025), too many people championing the use of 'Agile' in digital tech innovation still believe that moving fast and breaking things is the right way to eliminate poverty through digital tech (Terbeche and Carrier, 2019; Bisits Bullen, 2022; MzN International, 2023). The fundamental trouble with such a purist use of some aspects of Agile approaches is that the world's poorest and most marginalised have so little that they cannot afford for the little that they do have to be broken or experimented with.

The notion of Agile software development originated in the *Manifesto for Agile Software Development* proposed by 17 software developers in 2001 (Beck *et al.*, 2001). This had 12 core principles, focusing especially on customer satisfaction through continuous delivery of value, welcoming and harnessing change, short time scales, working together with customers, motivated individuals, sustainable development, simplicity, and reflective self-organising teams. While few of these were in themselves original ideas, bringing them together into a single manifesto was indeed seen as being innovative, especially in terms of the way that computer science had previously been taught and undertaken. 'Agile' has subsequently evolved with its own terminologies, reflecting its core notion of flexibility, and is best seen as 'an umbrella term for a set of frameworks and practices based on the values and principles expressed in the Manifesto' (Agile Alliance, n.d.). It is more than just frameworks such as Scrum, Kanban, and Extreme Programming alongside practices such as Pair Programming, Stand Ups, and Sprints, but frequently it is reduced by practitioners to a selective choice of these, particularly in the context of delivering development interventions.

Agile has served the software community quite well, but that does not necessarily mean that its principles should transfer effectively into other fields. Advocates from the private sector digital tech community who have been brought up in an Agile environment have nevertheless often implicitly sought to deliver digital tech for development through the mechanisms with which they are familiar, particularly through a combination of sprints, as opposed to what is often termed the 'waterfall' approach (traditional), as shown in Bisits Bullen's helpful framing in Figure 5.6.

In reality, project management in 'traditional international development' has never been quite as simple as shown in Figure 5.6, at least not as long as ICT4D or digital tech for development has existed as a more-or-less formal field since the late 1990s. In particular, regular ongoing monitoring and refining targets form an essential part of the log frame and theories of change approaches, which are often criticised as being within a waterfall framing, and have been widely used in development practice during this period. Moreover, both Agile and waterfall begin with the need to define problems, and all too often this is where the biggest challenge lies. Invariably, problem definition in both lies outside the communities that they

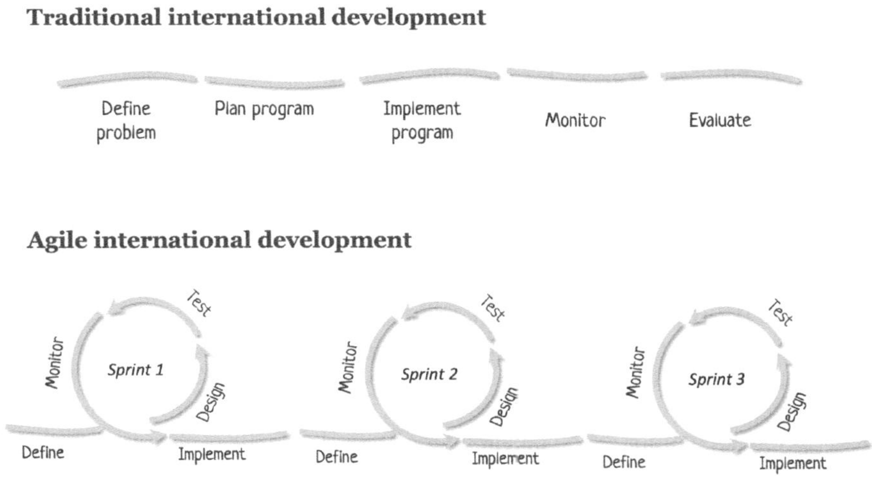

Figure 5.6 Bisits Bullen's (2022) framing of traditional and Agile international development

Source: Bisits Bullen, with permission.

are intended to benefit. They are instead shaped by external 'experts', the 'me's discussed in Chapter 4, and as typified, for example, by the McKinsey Global Institute's (2018) advocacy of Smart Cities.

In the international development context, strict Agile approaches suffer from three main problems. First, they are generally intended primarily for short-term delivery of a product, such as a piece of software, with sprints intended to last only for 2–4 weeks. While such short-termism may be helpful in projects only funded and designed to be completed within a year, it nevertheless still suffers from all the challenges associated with projectitis that were discussed in Chapter 4. Implementing an effective digital for development initiative requires all the skills of a marathon runner or steeplechaser, coping with the many hurdles and water jumps along the way, rather than those of a sprinter. Second, Agile project management focuses very much on designing and delivering 'for' customers, rather than 'with' them. Although, promoting 'customer collaboration' was a value behind the original Agile Manifesto, the word 'collaboration' interestingly does not actually appear in the 12 principles behind it. Instead, the manifesto states that 'business people and developers must work together daily throughout the project' (Beck *et al.*, 2001), and collaboration is in Agile practice used much more to refer to collaboration within and between self-organising teams (Agile Alliance, n.d.). Generally, it is not often intended that the customer should get involved in the actual development of the software by, for example, writing code. In contrast, good development practice should always focus on working together with the intended beneficiaries, helping them to achieve their ambitions, and giving them the skills to do so, with the intended outcome that they will subsequently be able to transform their own lives for the better in a sustainable manner. It is not doing it just 'for' the poor and then handing the completed product over to them, as in a turnkey project. Third, there is an unfortunate tendency for many Agile-led development interventions to adopt an 'experiment on' approach towards their intended beneficiaries. However, in good development practice, the 'product' is not just a piece of code that can be tested and experimented with, but rather human outcomes (see vignette by Juliette Unwin, p. 85). As noted above, Agile can work well for software development in novel circumstances, but the challenges facing the lives of the world's poorest and most marginalised are not novel. They are very well known, as are ways through which they can be improved, if only the 'innovation mob' were willing to read and learn from the experiences of past practitioners. In the short term, testing is usually of benefit more to the developers than it is to the beneficiaries, and if the tests go wrong, the implications for human life can be very damaging.

In practice, there is much to be said for an eclectic approach to delivering context-specific digital for development interventions and policies that draw on existing knowledge and good practices in project and programme management from many different perspectives. Agile has something to offer in terms of getting things done, especially in the context of innovation in the digital tech sector (Deardon *et al.*, 2010; Doerflinger and Deardon, 2013; Haikin, 2013; Dijkers *et al.*, 2018), but many of its most helpful aspects are not new and can be found in previous experiences and good practices in international development. Among these important shared values are the notions of flexibility, motivation, sustainability, keeping things simple, and ensuring self-reflection among all involved.

Learning from our mistakes

Innovation and neophilia often detract from the good of past experiences by placing excessive emphasis on the latest and 'best' new ideas and practices. Because AI is the latest digital tech to be promoted as being for 'good' development (see Figure 5.1), many of the lessons from experiments with blockchain, mobile devices, and thin-client systems, let alone distance learning in the early twentieth century, seem to be largely forgotten. Doing something new appears to be more important than doing something good, even when mistakes are made in so doing. This is wasteful of resources that can be ill-afforded in most development contexts, not only in terms of financial costs but also in the time taken to discover something for oneself that had already been discovered by others 20 years previously. No-one has the time to read the vast number of publications on ICT4D and digital for development that have been written since the late 1990s (see vignette by Sterling, p. 46), but it is a tragedy that so much good older material simply accumulates in digital repositories or on dusty bookshelves and remains unconsulted when a new generation reinvents the digital wheel.

This chapter began with the observation that innovation and failure go hand in hand. However, although we learn much from our own mistakes and failures, many of those involved in digital for development seem unwilling to learn from the mistakes of others, or even their own. An extensive evaluation of the World

Bank's activities in ICTs in 2011, for example, showed that only 50–60% of its work in ICT applications achieved their ICT-related objectives, and only 30% achieved their universal access objectives (IEG, 2011); very little was said about what the outcomes of any of their funding were in practice for people living in the countries where the initiatives were implemented. Such failure rates can quixotically be seen in a positive light if they are compared with Silicon Valley start-ups at the time, and the difficult contexts within which the Bank was trying to work (Vota, 2011). However, the real problem is that so few others learnt from the recommendations that came out of this report. Some in the ICT4D Community tried hard to promote learning from failures, as with the @FailFaire community on Twitter between 2009 and 2013, which attracted 617 followers, and with Fail Festival events in Washington, DC, and London between 2010 and 2017 (Vota, 2018). Although these were lively and interesting events, the appetite for such an explicit focus on failure has not been high among the wider communities of organisations implementing digital tech interventions in the international development context.

Four main reasons can be adduced as to why people do not learn enough from the mistakes of others (Aggarwal, 2018; Banks, 2013a; 2022). First, as previously discussed in Chapter 4, many of those educated in elite universities across the world have been brought up in a culture that makes them overconfident in their own abilities. They really do believe that they know best, and find it very hard to take criticism, even when it is well intended. They have no need to read and learn from the experiences of others, and the very idea of learning from 'failure' is anathema to them. Second, reading takes time, and in a world where many people choose to be bombarded by instantly forgettable social media posts and e-mails, they simply do not have the opportunity to read and learn from the detailed experiences of their predecessors. While AI may help in the processing of large amounts of material, as in the epigraph that introduces this chapter, it remains trapped in the past of its data and cannot yet convey the complex depth of detail required in delivering a successful intervention in a particular context. Third, the all-pervasive neophilia that dominates the digital tech sector means that success is largely defined through claiming that things are innovative and new, even when they are not. It is not just those making these claims who are the problem, but an even greater challenge is that their audiences, including those who fund 'innovation', are also frequently insufficiently informed and are therefore readily beguiled by the claims of novelty and the performance of the person delivering them. Fourth, it is no coincidence that the USA is among the top countries in terms of innovation (third in the Global Innovation Index in 2024; WIPO, 2024) and in percentage share of the digital tech sector (38% in 2024; Sherif, 2024). The 'American Dream', that everyone has the freedom and opportunity to succeed, is a powerful force driving people, and especially immigrants there, to feel that they must be successful, that digital start-ups are cool, and that they must never admit failure.

However, the reality is that most successful people do indeed learn from their own failures once they permit themselves to do so. The challenge is that this can be a long and painful process, and many people keep going on making the same mistakes over and over again because they cannot allow themselves to admit failure (Aggarwal, 2018). It is easy to say that they would not have made these mistakes if they had spent more time learning from others and reading more of the extensive literature on digital tech and development. However, this does not always seem to be the case, because the vast majority of digital interventions and initiatives continue to fail to serve the needs of the poorest and most marginalised as a result of the dangerous combination of ignorance and enthusiasm for innovation. Eskreis-Winkler and Fishbach (2019, p. 1733) emphasise the significance of ego in this, the 'me' syndrome again, when they write:

> Failure is ego threatening, which causes people to tune out. Participants learned less from personal failure than from personal success, yet they learned just as much from other people's failure as from others' success. Thus, when ego concerns are muted, people tune in and learn from failure.

Nevertheless, failure to learn from one's own experiences and from the experiences of others can be very expensive, especially for the most marginalised.

The costs of digital innovation failures for the most marginalised

Success and failure can be seen through the eyes of both the digital innovator and the wider communities. A digital entrepreneur committed to improving the lives of a particularly marginalised community could consider it a failure when their innovation did not achieve its altruistic objective, whereas an angel investor might see the outcomes of their innovation as a great success because of an unintended consequence that enabled the start-up to be highly profitable commercially. The words of the migrant in Malaysia cited in Harindranath's vignette (p. 96) in 2024 are worth repeating:

> Academic researchers and people from civil society organisations often come and ask us to answer surveys and questionnaires. But we never see them again. When they go away, nothing happens and our lives remain the same. When is this going to change?

This is a truism across the world, and applies not only to academic research, but also with many efforts of civil society organisations and companies working 'on' poor and marginalised communities. The poor and marginalised give so much to us; when are we going to reciprocate? They cannot afford our failures. All too often, digital innovations in international development seem to be undertaken for the sake of innovating, rather than for the real benefit of the most marginalised. These are the people who can least afford the theft of their identities, the loss of small amounts of money from mobile scams, their entrapment in online violence and sexual harassment (see vignette by Chimboza, p. 141), the ethnic abuse and cleansing through digital surveillance, and the appropriation of their jobs through digital systems.

Repetitive innovation largely for the sake of innovation and achieving competitive advantage (again the 'me' syndrome) rarely benefits the poorest, either at the policy level or practically. As Seth (2022) has highlighted so clearly, technologists need fundamentally to rethink their priorities and practices if the poorest are to benefit. Moreover, innovations can be used for both good and to cause harm, both deliberately and inadvertently through their unintended consequences. Research on developing autonomous systems for multiple co-ordinated UAVs (Piacentini *et al.*, 2019) that could benefit research and rescue in environmental disasters, for example, can equally be applied to battlefield or terrorist activities designed to kill large numbers of people. Scientists working in such fields have a difficult moral decision to make about whether they should indeed conduct such research, because ultimately they are as responsible for the resultant harms as they are for any good. It is no excuse to say simply, 'If I don't do it, others will.' Likewise, the introduction of national digital identities, as with India's Aadhaar system, Estonia's e-ID or Brazil's CPF digital ID (see vignette by Scur, p. 104) can provide benefits to all citizens but at the cost of greater surveillance and a new raft of potential digital harms, especially for the poorest and least tech-savvy (see vignettes by Al Dahdah, p. 121, and Rangaswamy, p. 131; and also Masiero and Shakthi, 2020; Panigrahi, 2021; Sadhya and Sahu, 2024; Keshvani-Ham and Tiwari, 2025). Invariably, it remains the rich and powerful who use digital tech to maintain or increase their power and control. The costs of digital innovation failure often hit the poorest the hardest. The digitally enabled theft of a small amount of money from a poor person will hit them much more than the loss of the same amount of money from a richer person. The use of AI systems based on data from the richer elites to shape policies supposedly 'for all' will privilege those elites and further marginalise the 'digitally invisible' poor.

The digital privatisation of India's administration

Marine Al Dahdah

(From Patna to Pondy, looking at health as a right, not a product)

The Indian Administration has long been portrayed as ineffective, corrupt, and uncaring. This is an image that the government has been trying to fix for the past 15 years, in part, through the national e-governance plan and the 'Digital India' programme that are contributing to the massive digitalisation of India's administration in the name of development. A cornerstone of this digital transformation, Common Service Centres (CSCs) seek to ensure the last mile of the Government-citizen relationship in every village of India (Figure 5.7). Using state-funded digital infrastructure – such as Aadhaar, UPI, eDistrict and eSevai platforms – these Internet centres are intended to serve as vital links between citizens and government services, providing access to public services.

Figure 5.7 Internet centre, Tamil Nadu, India, 2022
Source: Marine Al Dahdah, with permission.

However, this digital layer has made the State services increasingly expensive for millions of people, preventing them from accessing essential services such as healthcare, education and food. Unique point-of-sales for digitised State services as well as bottom-of-the-pyramid private services, CSCs became unavoidable fee-based intermediaries of administrative work. These centres have effectively privatised governance at the village level for profit. Their management and operations are not being held accountable to the public, and their primary focus is on generating revenue rather than serving the needs of people. Despite this, CSCs have become the face of local governance and are perceived by many elites as being efficient tech-savvy replacements of last-mile administrators who were under-equipped and undertrained to respond to the requirements of digitised administration. CSCs have disrupted the traditional channel of last-mile governance while creating new business opportunities for the multiple private companies they are tied up with.

Given the budgetary constraints weighing on local and central administration, with a spectacular decrease in civil servants and public investments at the village level all over India, it is indeed to be feared that the objective of digitised administration might ultimately be reduced to a fully privatised delegation to private local entrepreneurs for the purpose of saving money; eliminating completely the emancipating and 'empowering' rhetorics propounded by advocates of digital inclusion, leading to a lack of representation and input from local communities in decision-making processes, further exacerbating the disconnection between the state and villagers.

As noted at the beginning of this chapter, most innovations fail even against their own original objectives, which are in any case not usually designed to privilege the poorest. There therefore arises a fundamental question as to whether donors, international agencies, and foundations should be placing such a strong emphasis on innovation, especially for economic growth? Enough research and good practice have been built up over the last 25 years in the use of digital tech for development, that there is no excuse for such failure. Three important implications can be drawn from this assertion. First, serious consideration should be given by donor agencies to funding the wider roll-out of what is already known to work successfully, rather than funding digital innovation which is more likely both to fail and to privilege the richer more than the poorer. We know what works now; we need to get on and do it. Second, everyone involved should take responsibility for knowing the good practices that already exist, and focus on adapting them to the circumstances in which they find themselves. The innovation should be in the adaptation and repurposing of existing technologies for the poorest and most marginalised. Not knowing what has gone before should never be an excuse for innovative failure. Third, it must be recognised that innovation is difficult in large organisations such as UN agencies or global civil society organisations. Rather than being critical of this, it can instead be seen as an advantage, with their role being above all to share good practices, leaving the digital innovation to those in small, leaner, and more agile tech companies, social enterprises, and civil society organisations, who are mostly rich enough to afford to fail, should they choose the path of excessive innovation. This is a time of global restructuring driven largely by US citizens who elected Trump as their President in 2024, and it has resulted in very substantial reductions in Official Development Assistance (ODA) by most traditional donor countries, it has become ever more important to ensure that such assistance does indeed deliver on the needs of the poorest and most marginalised, rather than on the innovative aspirations of tech entrepreneurs and innovators.

There is, though, some ground for cautious optimism within the UN system. The Secretary-General's strategy on new technologies (UN, 2018, p. 4), while placing most emphasis on the 'incredible promise for the advancement of human welfare' of innovative technologies, does also recognise that 'They also hold the potential to generate more inequality and more violence.' The real innovation that is needed is a shift in the balance between these two tendencies. Inequality is only mentioned three times in the 20 pages of the strategy, and equity only twice. Those who claim that digital tech can improve the lives of everyone need to do very much more to mitigate the harms caused by innovation to enable their benefits to prevail. This will require the introduction of mechanisms to weigh up the potentials of innovations to be used for harm or for the good of the poorest and most marginalised. Scientists in universities and private sector corporations must never forget that they have individual responsibilities for the harms caused by their innovations alongside any good that they might provide. Put simply, they have choices: they can work to support and sustain the richer organisations and individuals, or they can seek to serve the poorest. In this context, it is useful to distinguish between hand-me-down innovation and pro-poor/marginalised innovation. We need much more of the latter (see vignette by Hughes, p. 115) than the former, but even sharing existing innovations more widely can still benefit the poor. Unfortunately, most digital for development initiatives are not even doing what we know would work best for the poor and marginalised, let alone promoting much innovation strictly for the most marginalised. The way that China has lifted half of its population out of poverty (Xi, 2014–2025; Fu *et al.*, 2021), and is currently fostering innovation explicitly to benefit its remaining poorest people, as with its focus on renewable energy development, and affordable devices, offers a challenging example to much US-dominated digital for development practice.

Evidence-driven decision-making in the use of digital technologies in education

David Hollow

(Co-Founder Jigsaw Education and Research Director EdTech Hub)

Building a culture of evidence-driven decision-making can help ensure that Educational Technologies (EdTech) are used to make a positive contribution towards addressing the global learning crisis. In 2008, I worked in Malawi and Ethiopia, conducting research on the impact of EdTech. During this time, the founder of the One Laptop per Child initiative suggested that my research was pointless. He said, confidently, that building evidence about EdTech in low-income countries was not necessary because the positive impact on learning was self-evident.

This mentality was indicative of many early-stage proponents of using digital technologies in education in low-income countries. Thankfully, significant progress has been made since then and there is growing global recognition that evidence should be central to decision-making (Akyeampong *et al.*, 2023; Hollow and Jefferies, 2022; Rodriguez-Segura, 2021). However, much work remains to be done. Too many EdTech programmes are still justified based on 'outputs', rather than 'outcomes' and entrenched economic and political incentives relating to EdTech often make evidence-driven approaches difficult to adopt (see Dercon, 2025).

This has significant implications for global education. The learning crisis is stark: in the world's poorest countries, nine out of ten children cannot read at the age of 10 (World Bank, 2022). There is a compelling social and economic case for addressing this crisis (Kaffenberger *et al.*, 2025) and yet funding for education is reducing significantly (UNESCO, 2025). Within this context, technology-based interventions are often presented as potential 'solutions', although the full financial costs are rarely measured accurately.

Influencing the detail of *how* technology is used in education is therefore crucial. The best way to do this effectively is by embedding practical evidence-building – through monitoring, evaluation, research and learning (MERL) – into all aspects of EdTech implementation and associated decision-making.

Everyone can contribute to building a culture of evidence-based decision-making in EdTech by asking the following question: *will this use of technology lead to an impact on learning outcomes that is cost-effective and works at scale?* Each part of this question is important and warrants unpacking.

- *An impact on learning outcomes*: this means ensuring that there is a demonstrable impact, over an appropriate time period, on the student learning outcomes related to the implementation. This should be based on statistically significant data that is generated using established learning assessment tools.
- *That is cost-effective*: this requires a transparent and consistent calculation of the implementation's cost-per-child (Mitchell and D'Rozario, 2022). This can be combined with learning outcomes data to undertake 'cost-effectiveness analysis' which enables comparison between implementations – a useful measure being 'Learning Adjusted Years of Schooling per USD 100' (Akyeampong *et al.*, 2023).
- *And works at scale*: the small-scale success of an implementation will not necessarily be sustained when working with a large number of learners across diverse contexts. Decision-makers need evidence of what works within the complexities of their own large-scale systems.

The practical outworking of this is inevitably complicated, contested and requires significant methodological depth (see Hollow and Jefferies, 2022). However, asking the question can serve as an initial catalyst to build better evidence, 'raise the bar' on our expectations of EdTech, and establish greater structural accountability for educational outcomes.

6 On freedom and digital enslavement

Tim Unwin

> As long as the problems of the poor are not radically resolved by rejecting the absolute autonomy of markets and financial speculation and by attacking the structural causes of inequality, no solution will be found for the world's problems or, for that matter, to any problems. Inequality is the root of social ills.
>
> (Pope Francis, 2013, #202)

If you cannot use digital technology for a day, let alone a week or a month, are you free? Are you not enslaved by the owners of the digital tech corporations whose hardware and software you buy and use, those I term the Digital Barons? As Pope Francis so eloquently and simply put it in the epigraph, it is essential to address the structural causes of inequality if the world's problems highlighted in the first part of this book, especially in Chapter 2, are to be addressed. This is a call for action, justified by the assertion that inequality lies at the heart of all social problems. The greatest form of inequality is to be found in slavery, when one group of people is owned in body and in soul by another group of people.

The notion of digital enslavement dates from the second half of the 2000s, when Rogerson and Rogerson (2007) presciently suggested:

> Digital slavery exists, unrecognised, unnoticed, unfettered, and, in many cases, unconsciously supported and promoted. Governments, for example, with the creation of biometric identification, have created the shackles of slavery. Digital beings are traded online as owners try to maximise market share and increase profit. Such trading is unacceptable and contravenes human rights.

The term has become more widely used during the first half of the 2020s with Chisnall (2020, p. 488) providing a particularly useful analysis of the issue, suggesting 'that the increasing trafficking of personal data to supply algorithm-based analytics and AI is enabling a new form of digital enslavement that has the potential to curtail liberty and cause harm' (see also Unwin, 2020b; 2022b). How we actively respond to such challenges has significant implications for the questions posed in Chapter 2 as to whether we should work towards using digital tech to *empower*

DOI: 10.4324/9781003598008-6

those who are being marginalised, or whether instead we should strive to create the conditions whereby we can all become *emancipated* from, and possibly through, digital tech, especially in the interests of the poorest and most marginalised in our societies. This chapter explores these issues by building on the discussion in Chapter 3 around rights and responsibilities, and then examines what being human means in terms of freedom and our engagement with the physical world in which we live (Nature). It argues that digital technologies are being used deliberately by companies and governments in ways that make citizens increasingly unfree, and that by entering into contracts with these companies we are becoming akin to indentured labourers in a new mode of digital virtual production controlled by the Digital Barons. The final section points to a dystopic cyborgAIn (combining cyborgs with AI) future for humanity unless we can free ourselves and others from the shackles of the slavery referred to by Rogerson and Rogerson (2007) above.

Balancing the individual with the communal: rights with responsibilities

Chapter 3 argued that the UN's human rights agenda must be balanced by a responsibilities approach through which it is made clear who is responsible for delivering and ensuring those rights, and that we as individuals have responsibilities to each other and thus to the various communities of which we form a part. Ideas about the 'common' and what is communal have evolved significantly over time. In much of medieval Europe, the feudal system provided for certain communal rights over areas of 'common land' that were intended for the mutual benefit of all those living in a specified place (Bloch, 1961; 1967). In the sixteenth and seventeenth centuries, a key shift took place when it was realised that enclosing these commons could enable individuals instead to invest productively in the land for greater profit, paving the way for the origins of European capitalism. Private ownership of land, replacing the earlier 'open' field systems, enabled emergent capitalists to extract surplus profits from the application of the labour power of freed people to these, once communal resources from which they were now excluded (see vignette by Fung, p. 136). It is essential for the arguments that follow later in this chapter to note the importance here of *free* people able to sell their *labour power* in return for a wage, but also to serve as a market with free choice over what they purchase. These two principles have underlain the subsequent development of capitalism ever since.

However, some communal resources have survived, mainly in the form of global commons, which under current international law include the oceans (and the seafloor bed), the atmosphere, Antarctica, and outer space (Nouwens, 2022). The problem here is that, although these might be 'common', they are not accessible to all. It is only the rich and powerful individuals and governments that can benefit directly from their use of these commons. The digital tech sector has been very effective at commandeering these commons in their own interest, most evidently in the use of outer space. In the 1960s and 1970s, it was only rich and powerful countries, such as the USA, Russia, and China, that had the expertise to launch satellites, but now many other smaller, but richer, countries as well as private individuals and companies, such as Musk (SpaceX) and Bezos (Blue Origin and Project Kuiper), are also able to exploit these commons in their own interests to extract even greater surplus profit. Most people have no possibility of accessing such notionally communal resources, and as a result are being further exploited through them by the privileged few who do. The same arguments apply to the extraction through seabed mining of the rare earth minerals that are so essential to current digital tech, but which only the very rich can afford to exploit (Crabbe *et al.*, 2024; Singh and Jaeckel, 2024). There are also those who claim that 'cyberspace', despite its mainly virtual rather than physical status, should be considered a global commons (Stang, 2013; Kaushik, 2021). Even were it to be treated as such, similar arguments would again apply, since the control of 'cyberspace' and the Internet remains in the hands of the few rather than the many who are increasingly exploited through it. In practice, cyberspace has been created by the rich and powerful with profit-seeking intent, rather than being part of the physical world of Nature which provided the basis for the original global commons – although there is increasingly a blurring between the physical and the virtual (for a wider discussion of Internet governance, see Souter, 2007; 2024; Hoxtell and Nonhoff, 2019; Denardis *et al.*, 2020).

This tension between individual and communal in the design and use of digital tech applies at all stages, from its inception to its ultimate beneficiaries and users. If a desire for individual benefit lies at the heart of digital design, then no amount of reversioning is ever likely to shift the balance towards communal interests (see vignette by Rowe, p. 23). Much Human Computer Interaction (HCI) design, for example, still starts in a research lab and begins with the individual researchers/designers who in the commercial world usually have a clear brief to work to. It is usually only when a prototype has been built that 'customers' are brought in to review a product, test it, and provide feedback (Rogers, 2012; GeeksforGeeks, 2024). Despite the rapid growth of emphasis on CX (Customer Experience; Quiñones and Rojas, 2023), very little HCI actually begins with a request from poor and marginalised individuals and communities to design digital tech specifically with and for them (see Hardindranath vignette, p. 97).

Despite this, there have always existed some communal and collective digital initiatives designed to try to serve communities rather than individuals. One of the most prominent of these was telecentre.org, founded in 2005 by Microsoft, IDRC and SDC, initially to provide communal access to digital tech to those who otherwise could not afford or have access to it (see also Chapter 4). At its launch, O'Neil (2005) commented that 'Telecentres provide a *community* gathering place, where people can use the Internet, learn new skills, and mobilize families and neighbours to address local social problems' (my emphasis). The subsequent changes in the tech sector, with its increased individualisation, made this original telecentre concept very difficult to realise in practice, and it is noteworthy that by 2025 telecentre.org's latest incarnation described itself as providing 'access to various technologies and solutions for *individuals* and organizations around the globe' (my emphasis; telecentre.org, 2025). This tension between communal rhetoric and practice on the ground is well captured in Rangaswamy's vignette (p. 131) where she asks

'Whatever happened to the community Internet?' in India (see also vignette by Tarmizi, p. 175).

The observation that communal digital initiatives were indeed promoted within the development sector in the early 2000s emphasises both how difficult it is to make these succeed, but also that many of the people involved in the early days of digital computing were indeed collaboratively and communally minded. This, though, was before the Digital Barons and the private sector more widely really grasped the commercial potential of digital tech, and how it could be used to reap great profits. The next sub-section therefore provides a brief overview of the FLOSS movements, Open Education Resources, and the exploitation of Open Data, all of which lay claim to having communal interests at their heart.

Social, shared and sustainable: Whatever happened to the community internet?

Nimmi Rangaswamy

(Professor, Kohli Centre on Intelligent Systems IIITH Hyderabad India)

At the turn of this millennium, India was abuzz with dreams of nation-wide digital connectivity – an aspiration that resonated well with the Millennium Development Goals, envisioning an Internet that was accessible, equitable, and safe for all, especially the marginalised. Fast forward to 2025, and we find ourselves amid a privatised, personalised, and pocket-sized Internet; a curated bouquet of services, packages, and products that vertically stratify access. The promise of inclusion remains, but the reality is clear – those who pay the most get the biggest slice of the digital pie. This short vignette pays particular attention to the mantra of shared access and connectivity, and the expanding telecommunication infrastructures offering a crucial boost to communities positioned along a spectrum – from the margins of poverty to the threshold of modernity (Best and Maclay 2002; Swaminathan and Kesavan 2016).

As of January 2025, 806 million people – 55.3% of India's population – were Internet users, leaving an estimated 652 million still offline. India now has over 820 million active users, with rising uptake in rural areas and a third of all users being women (Roy, 2024). Impressive figures – but do they reflect a community Internet that offers affordability, access, and digital literacy to the marginalised? In reality, the expansion has been largely market-driven. Growth is fuelled by private sector innovation and profit-seeking, rotating phone markets, proprietary ecosystems, and a broad spectrum of devices. A wide range of models – paired with flexible purchase options like EMIs (Equated Monthly Instalments), buy-backs, and bundled plans – has lowered the ownership barrier.

Yet, access remains uneven: Internet penetration is skewed toward wealthier populations, and data plans remain unaffordable for many. Meanwhile, usage patterns tell a story of privatisation: music and video streaming, social media, gaming, shopping, and news dominate, while job searches, upskilling, and citizen services are pushed to the margins (IAMIA (Internet and Mobile Association of India) and Kantar, 2023). What was once imagined as a shared, empowering infrastructure has become a personalised, pay-per-use playground – curated for binge and scroll rather than community or collective good. The Internet in India today is more a screen in the palm than a shared window. It entertains more than it empowers, and connects more to content than to community.

The Free and Open Software movements

One implication of placing more emphasis on 'we', rather than the 'me' discussed in Chapter 4, is that more collective methods of developing digital technologies and in decision-making about their use would need to be advocated. There is a long history of 'openness' within the digital tech sector dating back at least to the GNU (GNU's Not Unix) free software project announced by Stallman (rms) in 1983 to 're-evoke the once prevalent cooperative spirit among computer users' (Roberts, n.d.). This evolved into four main freedoms: to run the software (0), to study and change it (1), to redistribute it (2), and to distribute modified copies of it to others (3) (Free Software Foundation, n.d.). These were deeply embedded in philosophical and political notions of freedom and community, and the Free Software Movement gained considerable momentum when Torvalds released version 0.12 of the Linux operating system in 1992 under the GNU General Public Licence.

However, in 1998, following the release of the Netscape source code, the label Open Source was also coined as a more neutral, and less politically driven, term that would be acceptable to the business community, to refer to software that was superior because it was developed as part of an open development process engaging with potential users and developers (Open Source Initiative, 2006). More generally, the terms FOSS (Free and Open Source Software) and FLOSS (Free/Libre and Open Source Software to take the French language into consideration) were subsequently introduced to refer to both of these different, and competing, approaches to collective software development in the shared interests of different communities. This acted as a catalyst for software developers in some of the economically poorer parts of the world to focus on developing communal solutions through non-proprietary software, exemplified in Africa by the championing of FLOSS by Imfundo, among many others, in the early 2000s (see vignette on Imfundo, p. 11) and by the creation of FOSSFA (The Free Software and Open Source Foundation for Africa) in 2002 (Nwakanma, 2015). FLOSS nevertheless remains a minority interest, and the power of the major US, and increasingly Chinese, corporations to dominate these markets has been overwhelming (see vignette by Spiesberger, p. 134). The proprietary solutions developed by large corporations for their own profit, have significantly increased their power and control during the twenty-first century, despite the social, cultural, and political desirability of FLOSS software and coding practices. In February 2025, Linux thus only had 1.3% of the global operating system market, in contrast to Android's 40.5%, Microsoft's Windows (25.4%) and Apple's iOS (18.3%) (Statcounter, 2025). There are four main reasons for this relative lack of take-up of intuitively desirable FLOSS solutions in the interests of communal agendas: (1) the sums of money that private sector companies have to ensure their products and processes remain dominant; (2) the difficulty for young coders to forge a living based purely on self-exploitation and altruism; (3) the exploitation of 'Open' solutions by the private sector to enhance the productivity of their own proprietary software and apps; and (4) the notoriously poor ability of the FLOSS community to create novice-friendly guidance on using their products.

Open Educational Resources and Creative Commons

Similar challenges apply to the Open Content and Open Educational Resources (OER) movements that emerged in parallel with the Open Source movement in the late-1990s (Jhangiani and Biswas-Diener, 2017; see also Chapter 2 and vignette by Spiesberger, p. 134). These have become much more extensively used than has FLOSS, especially following the democratisation of the World Wide Web (WWW), and the launch of initiatives such as Wikipedia, and MIT's OpenCourseWare both founded in 2001 (Abelson *et al.*, 2021). The term OER was first coined in a UNESCO forum in 2002, and UNESCO has continued to champion their importance in supporting universal access to information, notably through their 2019 recommendations (UNESCO, 2022a; 2022b; see also COL, 2022). The Hewlett Foundation has also been strongly supportive, contributing US$170 million to the OER movement between 2001 and 2016 (Bliss and Smith, 2017). Integral to the development of OER was the establishment, also in 2001, of Creative Commons attributable licences based on the five rights to retain, reuse, revise, remix, and redistribute content. Although by no means all OER content is licensed in such a manner, it does provide an alternative model to traditional publication through proprietary licences (see also the work of the Commonwealth of Learning; West, 2009).

Making content freely available, or as OER through Creative Commons licences, nevertheless implies that authors are contributing the content because they believe they have responsibilities in the common sharing of knowledge more than they do in receiving monetary recompense. They are, in a sense, self-exploiting. Wikipedia, for example, drew on contributions from nearly 300,000 volunteers every month at the end of 2022 (Wikimedia Foundation, 2022). While creating OER may work effectively for authors who are university academics employed on good salaries, this is rarely an option for the world's poorest and most marginalised. Likewise, academics and others who share code openly through repositories such as GitHub, are frequently earning their main income from other sources. This is not a viable option for most of those living in the economically poorest countries of the world.

Spinning digital cotton to counter digital colonialism

Paul Spiesberger

(Dedicated to democracy, freedom, and universal human rights)

Massive data collection and online surveillance commenced in the 2000s as digital technology companies began to understand how advertising tailored to users' desires would further increase revenues. Online targeted advertising was established as a norm. Companies started to extract as much data about their users as possible to create psychological profiles better to match advertisements with users. With the support of governments and their security services, this led to a system of mass surveillance best described as surveillance capitalism (Zuboff, 2019).

Surveillance capitalism has led to the accumulation of vast amounts of data, which has most recently fuelled the creation of Artificial Intelligence (AI). By the mid-2020s, AI systems only became possible due to neural networks trained on huge data sets running on very expensive infrastructure consuming large amounts of resources (Strubell, Ganesh, and McCallum, 2020). The combination of data ownership and the very high costs initially made AI an exclusive business for just a handful of wealthy global private players. Others, such as universities, smaller companies, and states, are not able to compete with their AI models due to a lack of data and infrastructure. More recently, leaner Chinese AI systems do indeed offer alternatives to those developed by the major US corporations, but the basic principle of exclusivity of ownership and control remains sound.

The current development cycle of AI – extracting data, crafting/improving AI services, and then selling them back – creates a circle of dependence, which becomes more powerful with each iteration (Kwet, 2019). This has all been seen before, and is well exemplified by cloth manufacturing and the cotton slave trade within the British Empire in the 1800s. Resources were extracted (cotton/slaves), goods were manufactured (clothes/textiles), and monopolies were formed by crushing competitors with technological superiority (Banerjee, 2021). Mahatma Gandhi, in his struggle for Indian independence from the British Empire, broke the circle of dependence by spinning and weaving his own fabrics. He called upon others to follow in his fight for the self-governance of India (Brown, 2010). Today, the cotton is data and the clothes are AI or other digital services. Unless we find ways to spin our own digital cotton, we will continue to be ruled by digital colonialists. It is difficult to achieve prosperity for an entire nation if the institutions serve only an exploitative elite (Acemoglu and Robinson, 2013).

Policy remains the most effective means to challenge the current system, with measures such as breaking up monopolies and reforming the tax system to reduce the profitability of targeted advertising (Acemoglu and Johnson, 2024). While individuals can resist monopolistic control by adopting Free and Open Source Software (FOSS) and advocating for policy change, FOSS is often less convenient. Due to the lock-in effect, some individuals have no real choice and cannot easily switch.

Nonetheless, those who can should do the following:

- install a Linux operating system, use FOSS alternatives whenever possible, purchase repairable hardware you truly own, and opt solely for open file formats;

- support initiatives such as Public Money Public Code (https://publiccode.eu/) where publicly financed software developed for the public sector is made publicly available under a FOSS licence;
- encrypt data whenever possible and support privacy-focused initiatives;
- use decentralized Internet services, such as those built on the ActivityPub (https://www.w3.org/TR/activitypub/) protocol, which offer alternatives for social networking and can break monopolies;
- disrupt power structures by demanding policy change and start spinning your own digital cotton.

Open data and exploitation

Brin and Page in the late 1990s, with their original ambition for Google (n.d.) 'to organize the world's information and make it universally accessible and useful', had understood the power of knowledge and the potential of the Internet to be used to gain that power. They, along with other tech startup entrepreneurs, and especially the social media companies founded some years later, had hit on the brilliant trick that giving large numbers of people something of relatively low cost for free (such as searching, e-mail accounts, and social networking), created an opportunity for companies to gain vastly more valuable information about them all, not least as consumers, which could be monetised to generate great profits.

This process was considerably enhanced when these companies, along with mostly well-intentioned academics, started pushing governments to make their very large data sets available for free as Open Data (Kitchin, 2014). From the government perspective, the resultant innovation was seen as a way of driving economic growth, built in part on the academic research that they had often also funded. Meanwhile, academics were able to build their careers, and companies could make even larger profits based on free access to these valuable data and research findings. For most, this has been seen as a positive development, with the privacy and confidentiality issues too often being downplayed (Zuiderwijk and Janssen, 2014; although see Trepte and Masur, 2023; Benson *et al.*, 2024). Moreover, many such as Monino and Sedkaoui (2016) emphasise that data are available to everyone, and people anywhere on the planet can benefit. This, though, is misleading. Unless people are able to use these data in their own interests, they are unlikely to benefit directly. Most people in the world do not have the ability to access, let alone analyse these big open data sets, or conduct Open Science (UNESCO, 2023). As Monino and Sedkaoui (2016, p. xxix) comment, 'Exploited data generates billions of dollars' for the tech companies, but they do not go on to explore how the world's poorest can use these data to their advantage. Even the choice of the word 'exploited' emphasises the power relationships. It is not the data that are exploited, but rather the people whose data are used that are the source of the profit and who are being exploited.

Recent developments in AI and AGI (Artificial General Intelligence) offer much more significant challenges to the viability of OER and other forms of Open and freely available content such as Wikipedia. As Mackintosh (2025) has observed, 'Open Educational Resources (OER) are likely approaching an existential crisis because rapid advances in generative Large Language Models under the banner of "artificial intelligence" (AI) could erode their competitive advantage as digital public goods when compared to commercial content.' Artificial Intelligence is the logical outcome of the ravenously acquisitive, data-greedy digital corporations that have emerged in the wake of Google's original aspiration to have all the world's data in their hands, and then to monetise it. First in the firing line with respect to content and information is Wikipedia, despite the valiant defence of those such as Deckelmann (2023), who argues:

> The process of freely creating knowledge, of sharing it, and refining it over time, in public and with the help of hundreds of thousands of volunteers, has for 20 years fundamentally shaped Wikipedia and the many other Wikimedia projects. Wikipedia contains trustworthy, reliably sourced knowledge because it is created, debated, and curated by people.

The challenge will come from those who learn to trust AI more than they do people, and that is increasingly likely to come sooner rather than later.

In broad terms, Generative AI and Large Language Models (LLMs) pose five main challenges relating to the exploitation of content and reliability. First, there is a lack of transparency about the exact sources of content drawn on by each AI model, although a recent report by xfunnel (2025), drawing on 250,000 citations and 40,000 AI responses,suggests that 'domain authority, earned media, and user-generated content' are all used to varying extents. Second, there is widely reported bias within many AI models (Alvarez *et al.*, 2024; Ferrara, 2024), not least that these use little data from the poorest and most marginalised and can thus produce results that cannot be in the interests of these communities. Third, while AI tools can provide immediate and often free responses tailored to a specific refinable enquiry, users are also freely feeding information about themselves and their questions into these systems so that they can be improved; users are again exploited. Fourth, there are fundamental issues around licensing and copyright (Levendowski, 2018; Lim, 2023), with many authors and 'creatives' expressing serious concern about the ways in which the content that they have created is scraped up, unacknowledged and unpaid for, into AI. Finally, the energy requirements of AI systems place a heavy toll on all dimensions of the physical environment (Nature), and thereby add to the continually increasing environmental harms caused by digital technologies more generally (DESC, 2021; Crawford, 2024).

Learning from land rights so data rights are right from the get go

Mei Lin Fung

(People-centered leader advancing regional digital public infrastructure for MSME financing)

The painful history of poorly defined land rights, which in the past led to displacement through lack of formal documentation, offers a crucial lesson for the digital age. As people create online content, they often lack the means to claim ownership, leading to uncredited use and a drain on talent. To avoid repeating this exclusion at a greater scale, we urgently need digital public infrastructure with shared building blocks for identity, ownership, and traceability. This will ensure fairness, respect, and the ability for individuals and communities to protect and build upon their digital contributions, fostering meaningful participation for everyone by design.

I grew up in Singapore, where I saw how bold public investment in infrastructure and people enabled a small, resource-poor nation to become a vibrant, resilient society. That experience shaped my lifelong belief: that when people are equipped with tools to help themselves and treated with respect, the energy is unstoppable.

As the world gets more digitised, we must adapt this learning. Though the Internet has expanded access and opportunity, it has also brought to light fresh vulnerabilities and deepened divides. We must make sure people have the rights and means to benefit from what they create and are protected while doing so. People everywhere are creating digital content: music, stories, designs, innovations. Their contributions are copied, scraped, monetized – often without consent or credit. The talent pipeline is drained before it matures. Local entrepreneurial teams cannot get financing.

We urgently need *digital public infrastructure* built on shared digital building blocks – including identity, ownership, and traceability – to support participation with *fairness and respect*. These systems must be designed so that people can protect, reuse, and build upon their own contributions, individually and collectively.

We still have time to shape the digital future so that it reflects the dignity of everyone it touches – and ensures meaningful participation for *anyone, anywhere*.

Where do the responsibilities for digital lie?

The previous sub-section has highlighted how companies developing digital technologies over the past two decades have systematically exploited users, and particularly content creators, to reap rich dividends. In part, and paradoxically, this has been enabled by the gradual enslavement of previously 'free' individuals with personal rights, who have become ever more addicted to each iteration of new technology. Multiple individual freedoms are, in a sense, being exploited at a massive scale by a few individuals in mainly US and Chinese companies. This has significant implications for the earlier discussion around rights and responsibilities in Chapter 3, which questioned whether access to digital tech is a human right, and suggested instead that any rights-based agendas need to be balanced by a clear commitment by all sectors to a responsibilities framework (see Chapter 7). It is also important to reiterate that any right to be connected should also be balanced by a right to be unconnected.

Increasing recognition of the harmful impacts of digital technologies has nevertheless led to a growing number of civil society organisations at various scales, such as the Association for Progressive Communication, People-Centered Internet (see vignette by Fung, p. 136), the Just Net Coalition, the Open Rights Group, and the Alliance for Universal Digital Rights, advocating for digital rights, both human and legal, that seek to guarantee such claimed rights as privacy, freedom of expression, and access to the Internet. This is often on the grounds that these existing rights in the 'real' world should be maintained in the digital sphere (de Souza, 2025; Perry *et al.*, 2025; Roberts, 2025). The UN-led GDC (see also Chapter 3) thus emphasises:

> We commit to respect, protect and promote human rights in the digital space. We will uphold international human rights law throughout the life cycle of digital and emerging technologies so that users can safely benefit from digital technologies and are protected from violations, abuses and all forms of discrimination.
>
> (UN, 2024b, para. 22)

What the GDC failed to do, though, was specify exactly who had the responsibilities for doing so, other than that each stakeholder group had its own respective roles and responsibilities (paras 6, 8k, 22, 27), and a rather bland call for increased 'industry accountability' (para. 32b).

In practice, such well-intentioned right-based agendas have had less positive impact on the digital lives of the most marginalised than many had hoped for. Moreover, many rights-based initiatives ultimately benefit the digital tech companies more than they do the lives of the most marginalised. As at the origins of capitalism, these companies thrive on the ideology of freedom that means that consumers are free to use many such technologies and free to give away their identities and data from which the companies can reap profits. It is scarcely surprising that global digital corporations are so eager to support freedom-based initiatives, and to limit the powers of national governments to regulate the sector.

Little positive change will occur until a rigorous social, economic, political, and ideological responsibilities-based framework for the use of digital technologies is developed, laying out the responsibilities that governments, companies, civil society, and individuals all have to each other. If the poorest and most marginalised are to benefit, this involves changing the emphasis from 'me' to 'we' (Chapter 4) and prioritising an equity-based agenda instead of innovation-driven economic growth that benefits many, but not everyone (Chapters 2 and 5). Above all, it requires a shift from individual to communal thoughts, actions, and responsibilities. Chapter 7 takes forward these arguments for a practical responsibilities-based framework, drawing on the work of Walzer (1983), Williams (2006), and Waller (2011; 2014), and suggests that individuals as well as the three sectors (government, the private sector, and civil society), each has responsibilities to themselves, to others, and for certain actions.

Translating these arguments into the digital context, the fine sounding rhetoric of rights-based UN policies and agreements, such as the GDC (see Chapter 3), require a rigorously defined responsibilities framework, outlining exactly who from different sectors will be responsible for delivering which functions, if they are going to succeed. This will require very substantial negotiation and dialogue, and it is to be hoped that the UN Secretary-General and UNODET who have been tasked with producing the GDC Implementation Map, have the visionary capacity to ensure that all sectors and individuals, who endorse it, know exactly what their responsibilities are. Those who assert that meaningful connectivity to the Internet is now a human right, for example, must be clear about whose responsibility it is actually to provide it in any given context. If the poorest and most marginalised are to benefit from the use of digital tech as part of the wider community in which we all live at varying scales, we all have responsibilities for ensuring that they can do so should they wish. Invariably, this will require everyone to shift their emphasis away from the self and towards the other. Even a small shift can make a significant difference. The vignettes in this book have provided many examples of ways in which this can be achieved, and the concluding Chapter 7 draws these together, suggesting that the most important shift is one away from the self and towards the communal.

Freedom and Nature

Humans are social beings, and the creation of modern digital technologies has provided dramatically enhanced opportunities for people to engage in social networks at a scale that was unimaginable even in the 1990s. However, for many people this has come at the cost of disengagement with Nature, albeit that the idea of Nature is itself socially constructed, and thus it may be more accurate to refer to this as a reconfiguration of Nature. The use of digital tech mainly engages only two of our senses (sight and hearing); the crucial senses of smell, taste, and touch are mostly absent, despite ongoing research on haptics beginning to be used to introduce elements of the latter (Kuchenbecker et al., 2006; Qi *et al.*, 2024). We are constantly being persuaded that the virtual world is better than the real world, and yet we are becoming enslaved in a world that is only a partial representation of that reality. Our freedom to be truly human is being eroded, although for those who find difficulty in surviving in the real world, the virtual world can provide an apparently welcome, but destructive, escape from reality. All too often, this can develop into harmful addictions, and becomes merely an illusion of escape. Meng *et al.* (2022), for example, suggest that globally a quarter of the general population could be affected by at least one type of digital addiction, and that the burden is higher in low-income and lower-middle-income countries.

We are beings who flourish when fully immersed in, and engaged with, the physical environment in which we live (Nature). While any use of digital tech, in a sense, takes place in a physical environment, most of it also creates a barrier between the user and Nature through the use of a digital interface. Virtual Reality is lauded as being better than the real world. Even Augmented Reality (AR), while adding information about the places we are in, also separates us from them through the mediation provided by the technology. Recent estimates suggest that in six countries of the world (from a sample of 57: South Africa, Brazil, the Philippines, Colombia, Argentina, and Chile), users spend on average more than eight and a half hours a day online, with the figure reaching 9 hours 24 minutes in South Africa (Webcertain, 2024). In practice, most uses of digital tech separate the user from the world around them (see vignette by Chimboza, p. 141). In contrast, a wealth of research has shown that engaging in Nature has considerable benefits for human well-being, and the greater the diversity and quality of that engagement with Nature, the greater are the benefits (Mental Health Foundation, 2021; Ma *et al.*, 2023). Likewise, Ramadhan *et al.* (2024) conclude that 'digital detox interventions significantly reduce depressive symptoms, suggesting that intentional reduction or cessation of digital engagement may help alleviate contributing factors'. This suggests that digital separation from Nature may be one pathway through which mental health problems have become increasingly prevalent in many parts of the world over the last decade. Nevertheless, debate over the complex relationships between mental health and the use of digital technologies remains wide open, with many advocates from both sides being as much influenced by their ideological positions and research design as they are by any realities on the ground (Haidt and Allen, 2020; Santos *et al.*, 2023).

Increasing digital tech-related urbanisation across the world has also contributed to this growing separation between humans and Nature. World Bank (2024c) figures suggest that 57% of the world's population were urban in 2023, compared with 47% in 2000, and the greater use of digital tech by those living in towns and cities may be a further factor contributing to the digital separation of humans from the physicality of Nature. Throughout civilisation, there has been a tendency for urban elites to dominate those living in rural areas politically, socially, culturally, and economically (Potter and Unwin, 1989), and today is no different. Indeed, our notion of civilisation is itself fundamentally grounded in the urban, being derived from the Latin *civitas* (city) and *civilis* (civil). Urban elites dominating the digital tech sector are often living almost completely separately from the Nature that actually sustains urban life. Urban mentalities are consuming Nature for digital ends. This applies not only to the destructive extraction of rare earth minerals, but also in the increasing energy demands from each generation of new technologies, most of which is produced in rural areas. Chamanara *et al.* (2023) thus estimate that Bitcoin mining alone consumed 173.42 TWh of electricity between 2020 and 2021, which was more than the electricity consumption in most nation states. However, during the same period, its global water footprint was also about 165 km^3, which was greater than the domestic water use of 300 million people living in rural sub-Saharan Africa, and its land footprint was 1,870 km^2, which is larger than 26 of the world's countries. Artificial Intelligence probably has an even greater impact, with UNEP (2024) outlining the scale of potential harms and calling for the environmental impact of its full lifecycle to be assessed comprehensively. Although the efficiency of AI models has increased, and they can also be used to optimise energy production, they remain one of the main drivers of increases in the number and scale of data centres which the International Energy Agency (2025) estimates will account for one-tenth of global electricity demand growth by 2030.

Digital technologies can be used to provide a wealth of data on environmental change, to optimise energy use, to drive the circular economy, and to help develop more sustainable practices, but this comes at a cost to the physical environment and cultural landscapes. Despite an increasing amount of research on both the positive and negative effects of digital technology use on the environment, until recently there has been no rigorous holistic framework within which this complexity can

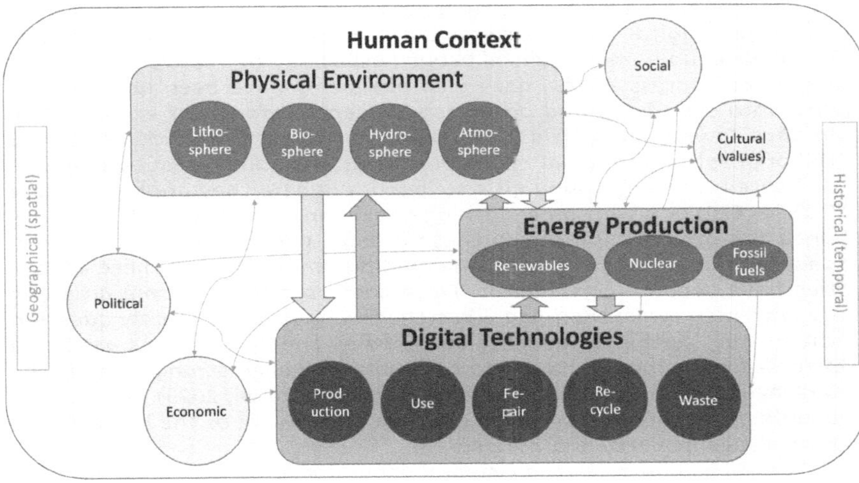

Figure 6.1 The DESC conceptual framework

Source: https://ict4d.org.uk/wp-content/uploads/2021/09/conceptual-framework-v3.pdf.

be assessed. One model that has been proposed is that by the Digital-Environment System Coalition (DESC), which is reproduced in Figure 6.1 (see also DESC, 2021). This emphasises the importance of assessing the impact of digital tech at all stages from its production to its waste on the lithosphere, biosphere, hydrosphere, and atmosphere, including outer space. Some of this impact is mediated through energy production of different types, including renewables, fossil fuels and nuclear, but much is direct. Furthermore, the human context (social, cultural, economic, and political) shapes and is shaped by these processes, and these vary both spatially and through time. Until recently, most global attention has been paid to the impact of digital tech on carbon emissions and waste, but Figure 6.1 emphasises that the processes involved are very much more complex and extensive than has often previously been considered.

The exploitation of young women: Digital tech at the heart of the immoral economy

Tendani Mulanga Chimboza

(Mentor, social change agent, leader, advocacy, innovator, mom, wife, villager)

Now more than ever, the excitement and attention of digital tech for development should be directed to the ethically sound use of technology. However, its significant role in enabling people to undermine social justice must never be forgotten.

The immoral economy thrives through exploitation and harmful practices. Digital technologies are a pinnacle of human development, but have equally marginalised people, especially in impoverished communities. Across Africa, dark web platforms have been used to lure girls into sex, drugs and other gender-centred forms of exploitation. In Cape Town (South Africa), for example, impoverished townships are full of people using social media platforms, including WhatsApp groups dedicated to luring young girls to 'parties' where they are paid a mere R 1500 (€70) or given some clothing in exchange for sex (Ruiters, 2023). Once the girls are lured into these groups, perpetrators flash money and luxuries to retain them. The chilling reality is that these activities have become normalised. From these platforms, some girls end up with men twice their age – but the family members rarely question the money, clothes or gadgets obtained by the girls. Moms are being arrested for using WhatsApp to communicate with strangers and selling their girls in exchange for money or food (Duval, 2025). Those who benefit from this immoral economy are well aware of the biggest vulnerabilities: poverty and hunger.

Digital tech enables the immoral economy to thrive in South Africa for three main reasons: first, social media applications do not have appropriate surveillance systems to flag and report potential abuse or exploitation; second, parents in impoverished communities lack sufficient digital literacy to monitor what their children are doing online; and third, South Africa lags in terms of proactive regulation, policy measures, or frameworks for safeguarding children online.

At a broader scale, South Africa has seen an increase of women being targeted by trafficking cartels since 2023. These women are lured through online platforms offering jobs abroad as far away as Thailand and Myanmar (DIRCO, 2025). Perpetrators leverage the naivety that persists in the minds of unemployed and frustrated young women in impoverished locations. The digital tech community cannot continue to turn a blind eye to this immoral digital exploitation of young women.

The Digital Barons and enslavement

The concept of Digital Barons has become increasingly widely used during the 2020s, popularised both by the notion of barons in the feudal era as in Technofeudalism (Varoufakis, 2023; Durand, 2024), and also by the industrialist Robber Barons of the nineteenth century in the USA (Klick, 2021). I was first drawn to the term in the late-2010s as being a useful general concept to capture the often unprincipled actions of a small number of individuals with great economic power, but I was also conscious of the challenges in applying a term usually associated with feudalism in the context of the twenty-first century (Unwin, 2014). In medieval feudalism, barons held fiefs in return for military and political services to an overlord, usually the king (Bloch, 1961, 1967). I therefore find it difficult to accept Varoufakis' (2023) notion that the contemporary owners of 'Big Tech' are a new form of feudal overlord fundamentally transforming capitalism. They do not offer oaths of fealty to a king or overlord, they do not perform military service in return for their fiefdoms, and the system of production through the labour of villeins on the manors of knights who hold these in return for military service to the barons (Maddicott, 1975), have few parallels in the contemporary world of digital tech. Varoufakis (2023, p. xiii) is, though, partly correct in his suggestion that, 'As for the rest of us, we have returned to our former status as serfs, contributing to the wealth and power of the new ruling class with our unpaid labour', although even here he fails to draw a sufficient distinction between villeins, serfs, and slaves. Capitalism emerged as a resolution of the contradictions within feudalism, and its internal logics are fundamentally different from those of its predecessor mode of production. In particular, in contemporary capitalism, there is no hierarchical structure of dues and services between villeins and manorial lords, between those lords and knights (often the same person), and between knights and barons, such as existed during European feudalism. The methods of surplus extraction are also fundamentally different, and as argued earlier (p. 129), depended on changed access to the means of production, often through enclosure, and the emergence of free workers who had to sell their labour power to the owners of land and the infrastructure of production.

Durand's (2024) thesis, inspired in part by his reading of Blankenship's (1990) account of a cyberpunk dystopia linked to techno-feudalism, revolves instead largely around a distinction between the coordination between the political and the economic. In his words, in the rollout of digital tech 'the imperfect but foundational distinction between the economic and the political realms in capitalism is increasingly blurred, a blurriness that was the norm in feudal times' (Durand, 2024, p. viii). He also places particular emphasis on the significance of violence and warfare in feudalism, which he suggests parallels many aspects of the contemporary political-economic nexus, as exemplified by the war against net neutrality (Durand, 2024, pp. 141, 143). While elements of feudalism may be evident in the contemporary political-economy of digital tech, this does not provide a sufficiently coherent justification for the idea that techno-feudalism has overthrown capitalism. Moreover, the extent and character of feudalism varied significantly both spatially and temporally in different parts of Europe, and indeed elsewhere in the world such as Japan and China (Chu, 2020), particularly with respect to the composition of the peasantry. To suggest that 'the spectres of feudalism are legion' (Durand, 2024, p. 148) is an exaggeration. Indeed, Durand (2024, p. 174) partially acknowledges this when he admits that 'The use of the idea of refeudalisation also requires a degree of caution.' All modes of production contain within them traces of their predecessors through the temporary resolution of the conflicts within them, but this does not necessarily imply that capitalism has been a failed project and our present is a return to an earlier logic of feudalism.

For these reasons, I prefer the use of the term Digital Barons much more in the sense of the Robber Barons of the USA in the nineteenth century. Hiltzik (2020) and Charles River Editors (2016) provide vivid accounts of the lives of the hugely wealthy and powerful magnates such as Rockefeller, Carnegie, Morgan, and Vanderbilt, and criticism of whose monopolistic power led to them being termed Robber Barons. Most of these men, though, were born poor, unlike today's Digital Barons, and made their wealth through controlling the railroads, shipping, steel production, oil and finance (see Figure 5.3 in Chapter 5). They lived the 'American Dream' to the full, championed capitalism, became staggeringly wealthy, and some subsequently contributed significantly to philanthropy, much as do today's leaders of the digital tech sector (see p. 173). Most were considered to be unscrupulous in their business dealings, and several, such as Jason (Jay) Gould (railroads and finance), were also apparently deeply unpleasant characters (Klein, 1997; Renehan, 2006). Most benefitted from the general absence of regulatory control, and used their political influence to the full to create monopolies, exploit labour, and engage in underhand financial dealings. Most of today's Digital Barons and the Robber Barons of the nineteenth and early twentieth centuries thus share six key characteristics in common:

1. They were unscrupulous.
2. They sought power (the 'me syndrome' at work again).
3. They competed among themselves.
4. They sought to circumvent rules and regulations.
5. They were male.
6. They lived in the USA.

Slavery has been with us universally for thousands of years, unlike European feudalism which flourished in a particular location and only for a limited period of time (Bradley *et al.*, 2011–2021). Slavery, however, has varied widely in its character and extent, from the Roman Empire around the Mediterranean, to pre-sixteenth-century Middle and South America, to the Islamic states of North and West Africa where it is estimated that a third of the population was slaves, to the slavery associated with the Mongol invasions and conquests of the thirteenth century, to the seventeenth-century Ottoman Empire where a fifth of the population of Constantinople were slaves, to the Triangular Trade whereby African slaves were taken to the Americas by Europeans, to China where slavery was only abolished in 1909 (Unwin, 2023c). Despite the rise of abolitionist movements in Europe in the mid-eighteenth century, slavery remains widespread today, and it is estimated that 50 million people across the world are now living in slavery (ILO, Walk Free and IOM, 2022). Slavery is generally considered to have two main forms, chattel slavery and indentured (or bonded) labour, but there are also various other forms of forced labour that can be included within the notion of slavery. It has coexisted, again to varying extents, with other dominant modes of production, including both feudalism and capitalism, and its extent has often coincided with periods of violent warfare during which conquered peoples were deliberately enslaved to provide labour. At its heart lies the notion that slaves are unfree, the property of another person to do with as their owner chooses. Although slaves existed within medieval European feudalism, they were not widespread, and in the light of the above discussion of techno-feudalism, it is important to distinguish between slaves and serfs, with the latter term, derived from the Latin *servus* (also translated as slave) often being used generically to refer to medieval peasants. Within Europe, though, it had a more specific meaning relating to land tenure within feudalism and the early modern economy. Whereas slaves were owned body and soul by a particular person, serfs were generally tied to the land of a manor in return for labour and other services due to the lord. Well-educated slaves could often be materially much better off than serfs. The archetypical form of serfdom in medieval Western Europe, especially England, was villeinage, through which villeins held land that they could cultivate for their own use in return for legally specified services, including labour on the lord's land. Other groups of serfs paid rent, worked as labourers for a manorial lord, or were among various categories of cottagers (Hilton, 1965; Hatcher, 1981).

Two critical points can be drawn from this brief overview: first, slavery was, and is, fundamentally a lack of personal freedom; and, second, slavery was not the defining form of labour in feudalism. This serves as another challenge to those arguing for techno-feudalism, but it also provides a starting point for a more detailed account of the ways that the Digital Barons and their coterie of acolytes are increasingly enslaving people in their thirst for wealth and power today (Rogerson and Rogerson, 2007; Chisnall, 2020). As the previous chapters of this book have argued, the world's population is becoming less and less free, as the major digital corporations, aided and abetted by complicit or naïve governments, increasingly entrap us in their webs. If we cannot spend a day offline, if we cannot pay cash to purchase food, if we cannot travel without being surveilled, if our health data are being sold for others' profit, if our creative works are being used to feed new AI models, or if we need to provide our biometric data to enter buildings, we are no longer truly free. However, this loss of freedom goes far beyond the superficial and affects the very essence of our being. Companies and governments are conspiring to make us give them our valuable data, our very identities, in exchange for relatively worthless and unnecessary benefits. They are the ones who benefit most from smart meters, health apps, the digital banking system, and digital identities. These processes are often based on unproven assumptions that digitalisation will provide cheaper and more reliable systems for companies and governments, because they require less human labour. Yet it is extremely difficult to calculate the precise savings, and few governments and companies are willing to share the exact costs of digitalisation, especially when they have been subject to ransomware attacks that add very significantly to the overall costs of the process. Invariably, poorly designed digital systems actually take much longer for users to negotiate than did previous systems in which real people provided good quality services. To be sure, well-designed, user-focused digital systems can indeed bring real benefits (see vignette by Proctor, p. 90), but this is all too often not so in reality. Moreover, the poorest and the most marginalised generally suffer most from such capture, often because they do not know how to use the technologies safely, wisely, and securely. They have much to gain, but also everything to lose. Increased digitalisation in essence usually passes the workload down to the consumer or citizen, who needs to learn how to use the system controlled by companies and governments and input all the relevant data. This is a real labour cost for the consumer-citizen. Moreover, it is all the more difficult to engage with these digital systems for those with disabilities, the elderly with dementia, the illiterate, and those from unsupported minority linguistic groups, who are often further excluded by processes associated with digitalisation (see vignette by Fiormonte, p. 40).

In contrast to previous epochs of conquest and physical violence being used to capture and control slaves, modern digital enslavement is much more nuanced and beguiling, usually taking the more subtle form of indentured labour. Agreeing to the terms of usage of digital platforms can, for example, be seen as paralleling the contracts or agreements between an indentured labourer and their employer, with there being little real choice for the labourer. Those who are constantly on social media have thus been persuaded voluntarily to provide their data (a labour process) to the Digital Barons, who then manipulate and exploit it through the infrastructures they have created and own, to generate surplus profit. Meta thus had some 3.35 billion people working for it in 2024 (all those using its platforms), with most of them not consciously realising they are actually generating its profits, and few receiving any wages for their labour. They certainly receive what they perceive to be some benefit in kind ('freely' sharing and reading posts or watching videos online), but this is itself a further loss of freedom, akin to nineteenth-century mill workers having to buy produce from the owner's mill shop through tokens rather than cash. A cash wage would instead provide labour with freedom of choice over how that money can be spent. People do not receive credits from Amazon for using Google products. Furthermore, most consumers and citizens do not have the wealth or the knowledge to acquire these means of production for themselves, and so are further excluded from the wealth acquisition processes put in place by the owners of these companies. At the same time, increasing numbers of people using digital technologies, and especially social media, search engines, and AI, considerably increase the potential market size for online businesses capable of benefitting through the analytical use of these technologies. Although influencers may appear to the world's poor to be gaining unimagined wealth from their posts on social media (Hund, 2023), they too are but pawns in the wider process of digital enslavement in the interests of the Digital Barons.

These arguments have significant implications for the question asked earlier in this book (p. 21) as to whether marginalised people and communities can indeed be empowered through the use of digital technologies, or whether like the serfs of nineteenth-century Russia their only future lies in emancipation? If, as has been argued above, we are all becoming increasingly enslaved through our use of digital technologies, then the logical conclusion for those who aspire to freedom is that digital technologies in their present form will never be able to be used to *empower* the most marginalised. Rather, the Digital Barons are seeking further to enslave them through forcing them to use these technologies (see vignette by Spiesberger, p. 134) in their polydipsia for ever larger markets and cheaper labour. In this sense, those who are fortunate enough not to be connected may well be the lucky ones, the only ones who can still engage freely in Nature. However, even that opportunity is being further eroded as the use of digital technologies increasingly also destroys the physical environment in which we have our being. If this is so, then focusing on *emancipatory* processes and practices may be the only way through which we can break free from the shackles of digital slavery. The lessons of the 1861 Edict of Emancipation across the Russian Empire and the subsequent revolution of 1917 nevertheless also need to be heeded by those eager to ensure a peaceful transition to a new and better world.

Towards a cyborg future-present

The idea of humans becoming increasingly enslaved through the use of digital tech is concerning enough. However, one of its implications is that increasingly sophisticated cyborgs are likely to be replacing many human activities very much sooner than was once envisaged. Digital technologies can already be used to do most things that humans can, but do them better, more reliably, more accurately, and faster. Some, such as Blum and Blum (2021) with their Conscious Turing Machine, envisage that machine consciousness will be a reality in the near future, although others consider this less likely (Oliveira, 2022). Much depends on how consciousness is defined. The first humanoid robot boxing match took place in May 2025 (Weilan, 2025), and AI-powered female sex robots (sexbots) have become increasingly sophisticated since the first iterations evolved from earlier dolls in the 2010s (Devlin, 2018). The US military has been actively involved in exploring the implications of human/machine fusion for the creation of cyborg soldiers by 2050 (Emanuel *et al.*, 2019). Whatever one thinks of these examples, the reality is that cyborgs are already with us in many different guises, and they are rapidly becoming increasingly sophisticated.

The world of digital tech has changed fundamentally since Clynes and Kline (1960) first proposed the word 'cyborg' to refer to cybernetic organisms as human-machine systems that could survive in extraterrestrial environments, and Haraway (1991, pp. 150–151) warned that 'the cyborg is also the awful apocalyptic *telos* of the "West's" escalating domination of abstract individuation, an ultimate self untied at last from all dependency' (see also Hamilton, 1997). Two broad streams of enquiry have emerged since then: the practical and the conceptual/philosophical, and there is often insufficient interaction between them. Engineers insufficiently address the ethics and cultural implications of their actions. Practical experiments at creating human-machines have evolved rapidly, ranging from biochips in humans, to the construction of new types of prosthetics, to brain-computer interfaces (BCIs), the creation of cyborg tissues, and extending life through the use of digital tech, typified by the biotech work of Alphabet's Calico Labs. Conceptually, many different perspectives on cyborgs have been proposed, ranging from those who have built on some of Haraway's (1991) ideas and consider that all our existing interactions with digital tech can imply that we are already cyborgs, to those who wish to retain a clear physical distinction between machines, humans, and machine-humans or human-machines. Tsvetkova *et al.* (2024, p. 1864) thus observe that 'From fake social media accounts and generative artificial intelligence chatbots to trading algorithms and self-driving vehicles, robots, bots and algorithms are proliferating and permeating our communication channels, social interactions, economic transactions and transportation arteries.' They go on to make the important point that 'Networks of multiple interdependent and interacting humans and intelligent machines constitute complex social systems for which the collective outcomes cannot be deduced from either human or machine behaviour alone.'

Increasingly, therefore, we need to develop completely new ways of understanding this cyborg interface. Some of the most interesting ideas have emerged within the domains of law, ethics, and policy, where discussions have been ongoing for well over a decade. Among the most prescient of these was by Wittes and Chong (2014, pp. 14–15), who usefully noted:

> Until we know how far down the cyborg spectrum how many of us are going to travel, it is folly to imagine that we can fashion definitive policy for surveillance – or anything else – in a world of cyborgs. It is more plausible, however, to imagine that we might discern certain principles and considerations that should inform policy as we mature into adolescent cyborgs and ultimately into adult cyborgs.

They highlight three main arenas where urgent policy discourse and agreement are still required a decade later: data generation, data collection, and constructive integration. As they ominously concluded:

> The answers to these questions will not always be the same. We will not feel the same way about the privacy of what you see with your bionic eye – and your right not to incriminate yourself with the images it collects – as we will feel about your right to shield physiological data you collect on yourself recreationally or to incentivize your own fitness.
>
> (Wittes and Chong, 2014, p. 21)

The implications of our increasingly cyborg-influenced/controlled present and future for the lives of the world's poorest and most marginalised receive insufficient attention both in practice and in theory. Certainly, developments at the interface of robotics and AI are already moving from the experimental to the commercial in richer parts of the world, typified, for example, by the robotics displays at the UN ITU's AI for Good summits (Figure 6.2). There are some who also envisage the potential benefits that this may have for disadvantaged groups, particularly with enhanced prosthetics and support for people with dementia. However, these come at a financial cost that is beyond all sense of reality for those in economically poorer parts of the world, and thus once again provide a vivid example of digital tech further exaggerating inequalities at all scales.

The wider ramifications of our increasingly cyborgAIn (emphasising the combination of cyborgs with AI) future are of fundamental importance for the lives of the poor and marginalised. This is regardless of whether we consider using a mobile device makes us a cyborg, or whether we choose to restrict such terminology to the physical enhancement of human bodies through implants or attachments. If cyborgAIns are already better at doing many, or even most, traditionally human functions than humans are, this has deep implications for the most marginalised, particularly relating to labour, the provision of services, and

Figure 6.2 Robotics displays at the UN ITU's AI for Good Summits in Geneva, 2024 and 2025

Source: Tim Unwin.

relationships with Nature. There is already a labour crisis in many of the world's poorest countries, particularly in Africa, where a combination of poor/irrelevant education and a lack of employment opportunities means that millions of people cannot earn a viable income and thus still live in poverty; 692 million people globally live below the $2.15 per day poverty line in 2024, and eight countries have more than 60% of their population living in poverty (World Bank, 2024d). A future where cyborgs perform most productive economic functions puts the labour of the world's poorest at even greater risk, and it is difficult in such a scenario to envisage any 'useful' purpose for most human labour, especially at the margins of the productive economy. Moreover, government and privatised public service functions will also largely be delivered by cyborgAIns, usually 'trained' on data from the richer ableist parts of the world, which will further control and disadvantage the most marginalised. Furthermore, being forced to engage and interact with cyborgAIns, will separate everyone yet further from the physical environment of Nature that is ultimately the source of our sustained existence. The consequent dislocation and disorientation are likely to make the lives of the poorest yet further untenable, with potentially catastrophic socio-political ramifications (see also Yudkowsky and Soares, 2025).

It has until recently generally been argued that cyborgs are basically human with some machine attachments. However, if machines are already able to do many things at least as well as humans, it seems likely that the cyborgAIns of the future will actually be machines with a small amount of whatever humans can still do better incorporated within them. Over time, new generations will include less and less that is human. Identifying this human essence, both physically and conceptually, becomes one of the most important tasks for future generations of scientists and philosophers. Nevertheless, following the arguments of the previous section, it seems probable in the short term that digital enslavement will become ever more complete, with humans being 'persuaded' in increasingly subtle ways to indenture themselves to a small number of global corporations, mainly headquartered in China and the USA, that at the same time also benefit from states requiring all citizens to participate through means such as digital identities and open data. Data will increasingly be extracted from human lived experiences to control and surveil our behaviours in the interests primarily of maximising profit and state security. As Pope Francis asserted in the epigraph at the beginning of this chapter, if we wish to change this, we need to reject the absolute autonomy of markets and financial speculation. Although our choices will become increasingly predictable and constrained, humans will still need to play a role in consumption for such profits to be realised. It is perhaps here, in the apparent irrationality and randomness of individuals that a space for freedom in the increasingly competitive world between the behemoths of the Digital Barons may still be found. It is also in the genetic diversity referred to previously (p. 72), that a truly human future may remain, despite increasing attempts to edit out what are deemed by medical scientists to be 'faulty genes'. The Digital Barons aspire to make perfect cyborgAIns, when we ourselves are diverse and not uniformly perfect.

This dystopic vision of the future is one that pays little heed to the lives of the world's poorest and most marginalised. It is a world very far removed from feudalism, or indeed from Lenin's (1917) notion of imperialism as the highest stage of capitalism. Whether it evolves fully into a completely new mode of production remains to be seen. However, it raises important concerns around the fundamental question of who benefits most from the changes occurring in the development of digital technologies, and what those committed to equity and the reduction of global and local inequalities need to do to change the course of history. It is to these that the concluding chapter, Chapter 7, now turns, offering some bright optimism to balance this chapter's dark pessimism.

Beyond the cable: 'The embrace of co-designed, plural futures'

Tom Wambeke

(UN executive advancing learning innovation and digital strategies for inclusion)

Digital inclusion is often reduced to the task of connecting the unconnected – as if access to devices, networks, and bandwidth alone could bridge a divide. This narrow, tech-centric perspective reflects a form of linear technological determinism that overlooks the intricate, context-specific realities experienced by billions worldwide. Defining inclusion merely in terms of infrastructure and digital tools misses the deeper, human dimensions that give technology its significance. In many regions, fostering inclusion means more than enabling connectivity; it requires addressing entrenched layers of exclusion, historical trauma, and systemic marginalization – challenges that technology by itself cannot resolve.

This is why we must stretch the boundaries of our imagination and craft new narratives grounded in *simplexity* – the ability to engage with complexity without flattening it. True digital inclusion cannot emerge from one-size-fits-all solutions dropped into communities whose lived experiences, languages, and local histories diverge significantly from those of the designers. Rather than framing inclusion as a matter of technical fit or market reach, we need to reconceive it as a cultural, political, and ecological endeavour – one that resists the urge to abstract human experience into reductive categories like 'user needs'. Embracing complexity is not a barrier to progress; it is the path to just, and context-sensitive inclusion.

The philosopher Ludwig Wittgenstein (2023) once remarked, 'Whereof one cannot speak, thereof one must be silent.' This sentiment resonates deeply with the current limitations of the digital inclusion discourse – a space often saturated with confident proclamations, yet lacking the nuance to capture its full human complexity. It was from this recognition that *Posterheroes 2021* emerged: a global visual storytelling initiative inviting artists to reimagine what digital inclusion could mean across diverse contexts (ITCILO, 2021). One of these images is reproduced here (Figure 6.3). Rejecting the clichéd imagery of glowing screens and orbiting satellites, these artists offered quieter, more grounded visions – scenes of interdependence, resilience, and everyday struggle. Their work stands as a visual counter-narrative to techno-solutionist aesthetics, reminding us that genuine inclusion is not something imposed upon communities, but should be co-created with them, rooted in their stories, textures, and aspirations.

The call, then, is to move beyond inclusion as mere access toward inclusion as *transformation* – a redefinition that gestures toward a truly radical inclusion. This vision demands co-designing for the

Figure 6.3 Remember what it means to be digitally literate today may change tomorrow

Source: ITCILO (2021) reproduced with permission.

pluriverse – a world where multiple ways of being and knowing co-exist. In such a world, digital tools are not delivered as symbols of progress, but cultivated as instruments of solidarity, storytelling, and survival. It means elevating disabled voices, indigenous epistemologies, gendered experiences, and informal economies – not as peripheral exceptions, but as essential sources of wisdom. It also requires humility: recognizing that inclusion may sometimes mean safeguarding the right to remain disconnected, autonomous, and unseen.

By embracing this stance, we resist the ongoing privatization and control of the digital commons by corporate and geopolitical powers, and instead reclaim it as a space for shared creativity, care, and collective becoming. Here, inclusion becomes a verb – a practice, not a performance. It is enacted through deep listening, critical reflexivity, and a commitment to designing as if people – not platforms – truly matter.

7 Towards emancipatory responsibilities and actions

Tim Unwin

> We could list the facts of world hunger, poverty and famine endlessly. But facts alone do not tell us what to do. What surely matters is action. Which action we advocate depends partly on our perception of the facts, and this perception depends partly on the ethical outlook we adopt. Both our perception of problems and our prescriptions for action reflect our ethical theory. Ethical theories are not elegant trimmings that decorate our reasoning about practical problems. They determine our entire focus. They lead us to see certain facts and principles as salient and others as insubstantial. They focus our action – or our inertia.
>
> (O'Neill, 2016, p. 30)

This book has outlined what I see as being the five main reasons why digital technologies have not sufficiently been used to serve the interests of the world's poorest and most marginalised:

1. an emphasis on economic growth rather than equity;
2. the failures of the UN system;
3. the 'me' syndrome;
4. the innovation fetish;
5. digital enslavement.

Many of these challenges are also relevant more widely in 'international development', but they are writ large and exacerbated through the use of digital tech. The emphasis has been fundamentally on the needs and interests of the most marginalised. These are the first billion in terms of any socially and culturally responsible development agenda; those who should require our greatest attention. They are not pejoratively the 'Bottom Billion' (as termed by Collier, 2007; and in Onyango's Forgotten Bottom Millions, 4BM, 2025). Language matters. However, as O'Neill (2016) reminds us in the epigraph, we need to go beyond just describing and explaining what we observe, and move urgently towards *action*, based upon an ethical vision of what should be, rather than just what is.

DOI: 10.4324/9781003598008-7

The arguments of the previous chapters suggest above all else that digital tech cannot be used to empower the world's poor, because those who create and promote these technologies do so with an interest in exploiting those poorer than they are. That which is created to enslave us and take our freedom away cannot be empowering. The question raised in Chapter 2 (p. 22) thus appears resolved in favour of the need to seek emancipatory actions that will enable the world's poorest and most marginalised to benefit (Horkheimer, 1972; Habermas, 1974; 1978; Gross, 2010; Unwin, 2017a). Again, as emphasised in Chapter 2, such emancipatory actions lie not just in the hands of the poor. All those engaged in the field of digital tech and development, whether we work in private sector companies, governments, or civil society, need to become emancipated from the illusions of our own captivity. We can only do this once the scales have been taken from our eyes, and the myths are shown to be the fantasies they are. This book has sought to shed light on the structuring of these mythologies to help develop an emancipatory understanding of the reasons why digital tech as it is currently deployed increases inequalities, and insufficiently addresses the needs and aspirations of the poorest and most marginalised. It is thus truly part of an enlightenment project, and rejects the nihilism of post-modernity. As so many of the vignettes have suggested, it should be about 'we' rather than 'me'. It is about our individual and collective responsibilities to each other, and it is about learning from the past rather than repeating the mistakes of previous 'innovative' digital interventions.

Merely changing our vocabulary is not enough

There is an easy answer to the many challenges posed by this book. We could simply admit that digital tech is indeed used to increase inequalities, and has not and will not under present conditions be used to empower the poorest and most marginalised. We could be honest and focus on the easy next billion rather than the more difficult 'first billion'. This could have been the starting point for the Global Digital Compact (GDC), along with recognition of the very significant harms caused through the use of digital technologies. Beginning with this observation clearly in mind would have led to a completely different framing of the UN's agenda for the future, one that would not have served the powerful interests behind the document that was agreed in 2024. To be sure, there is some instrumental recognition later in the GDC that digital tech can be harmful, but this is not there at the very beginning of the document; it does not underlie the text that was agreed. It is not grounded in an ethical position, as called for by O'Neill (2016). Likewise, while the harms of using digital tech are indeed beginning to be recognised in wider UN statements and documentation, all too often this appears to be as a reluctant afterthought, and not centrally placed as a fundamental characteristic of the use of digital tech in development. The dominant global rhetoric around technology and development remains one of belief in technological determinism and digital instrumentalism, and that this does good.

It is certainly difficult politically to acknowledge that the use of digital tech increases inequalities and that the most marginalised will never truly benefit from it because the so-called 'digital divide' will always get bigger. However, this is the reality, and this is the main problem that needs addressing if we wish to be honest about digital tech and development. The more marginalised and isolated a person or community, the more difficult and expensive it becomes to serve them, not least in terms of digital connectivity. Companies and governments therefore need to be brave enough to say that at present the most marginalised people, those living in isolated mountain villages or pastoral nomads traversing semi-arid lands, refugees fleeing oppression in their homelands, women and girls who are restricted by men and boys in their behaviours in strongly patriarchal societies, ethnic minorities who cannot read the languages of the Web, as well as those with disabilities who cannot engage with much digital tech, will not benefit because we cannot yet afford to provide them with appropriate and secure digital technologies.

However, if we do admit that it is not possible to ensure an 'inclusive digital future for everyone' (IGF, 2024), we must then ask what percentage of the population are we going to ignore and say we cannot afford to deliver for them? Chapter 1 emphasised that in 2024 one-third of the world's population did not use the Internet, and around half of all rural people likewise did not do so (ITU, 2024). Worryingly, the curve of overall Internet use is also now levelling off after a period of growth between 2017 and 2020. If universality (ITU, 2022; see also Chapter 3) is not an option, what figure might the UN and governments opt for? At current rates, it is likely that 25% of the world's population will still not be connected to the Internet in 2030; optimistically, this might just reach 20%.

The apparently easy answer is after all, not so easy. Why is it then that the coalition of interests behind 'universal and meaningful connectivity' (ITU, n.d.) are not willing to let go and admit the truth? In part, it is because of the inertia about which O'Neill (2016) warns in the epigraph. In part, though, it is also because the private sector is desperate to increase its markets, and enslave more of the world's population in the pursuit of economic growth. An increase of 10% in the number of people connected represents about 604 million new people whose data could be mined and exploited. In part, it is also because it is politically unacceptable in a world dominated by human rights agendas to exclude people from what is widely seen as being beneficial, or indeed a right. However, this logic changes fundamentally if the unconnected are instead seen as being the truly fortunate ones, still benefitting from their freedoms. A moral choice has to be made. Those who strive for universal connectivity believe that it is 'good', but for those who value freedom in nature above all else, perhaps it is not so good after all; perhaps the AI for Good summit should be renamed AI for Bad? (see Figure 5.1 in Chapter 5) – but then it would not attract the levels of funding that make it such a glamorous and enticing event.

Responsibilities and emancipatory interests

The previous section serves as a reminder of the fundamental importance of a moral compass and a consideration of ethics in any discussion of digital tech for development (Unwin, 2009; 2017a; Fuchs, 2023). These provide the basis for the action that O'Neill (2016) has so clearly articulated in the epigraph. However, as she also notes, inertia is a choice, and many of those involved in digital tech for development made that choice long ago, some undoubtedly unconsciously but many consciously (see Chapter 4). I retain the hope that most of those working in the field of digital tech for development have within them at least a flicker of aspiration that they are doing wider good of some kind. This book has the intent of fanning that flicker into a powerful flame, to reinvigorate all our efforts to work *for* and *with* the poorest and most marginalised to mitigate the harms and ensure that they do indeed benefit from the potential of using digital technologies to improve their lives. While these arguments have been grounded primarily in my own understandings of Critical Theory, many different philosophical positions concerning social justice would point in a similar direction (see, for example, the differing views of Rawls, 1971; 2001; Sen, 1992; Sandel, 1998). This agenda does not, though, necessarily require a primary focus on justice and obligations. It is relevant to anyone who adheres to values and morals that not only emphasise empathy with the poor and marginalised, but also expects us to have responsibilities as individuals and collectively to help them improve their lives and thereby make the world a better place (see vignette by Tarmizi, p. 175). In so doing, we must nevertheless recognise that our offers of assistance may well be rejected. Once again, the right to be unconnected should be treated as equally important as any right to be connected.

I draw a key distinction between obligations and responsibilities. The former are in essence externally imposed legal requirements, whereas the latter are internalised moral choices reflecting values and usually requiring some degree of freedom of choice (see also Haji, 2023). The two are clearly related, but I suggest that it is necessary to have a thorough understanding of individual and collective responsibilities before proceeding to develop any formalised systems of legal obligations that might be imposed on us (O'Neill, 2016) Specific rules and legal requirements about using digital tech also ultimately need to be in place through international agreements, as in managing the radio spectrum, but, as with human rights, such agreements do not always preclude 'bad actors' simply choosing to disobey or circumvent them. Moreover, what some see as the collapse of multilateralism and the rules-based international order, particularly following Trump's unilateral efforts to 'make America great again' (Patrick, 2025), could reinforce the need for an alternative approach to such agreements that is based more on responsibilities than on obligations, and draws on a wider spectrum of international interests. I want to retain the possibility that there is enough shared moral agreement to shape the future use of digital tech in the interests of the marginalised, even though the golden rule cannot provide the basis for such a possibility: 'I' do not necessarily know what 'you' would like me to do for or to 'you' (see Chapter 3). Such an argument is not necessarily based on a theoretical sense of justice, but rather on how we as individuals react to the moral challenges of the processes of marginalisation.

Figure 7.1 of a young girl holding the hand of a blind boy begging from rich people in a car in the 1970s serves as an analogy. The car is a piece of then modern technology, not dissimilar in practice to digital tech today. Most readers of this

Figure 7.1 Children begging, West Singhbhum District, India, 1977
Source: Tim Unwin.

book will be similar to the (invisible) people in the car. The question remains, 'what should we do?'. We can drive off in our latest hyped-up technological innovation, or we could change course and work with the children to help them develop new digital interventions that might be used to improve their lives. We all need to become emancipated, freed from our illusions about digital tech and the shackles that constrain us, so that we are able to help others in turn to achieve their own emancipation.

This requires a broader consideration of responsibilities, building on the comments in Chapters 3 and 6, where the case was made that the Human Rights agenda needs to be balanced by a responsibilities agenda, that generic mentions of multistakeholder responsibilities in UN documents, such as the GDC, need to be made much more specific, and that we need to recraft understandings of the value of communal responsibilities in serving the interests of the poorest and most marginalised. Waller (2011) has argued that moral responsibility cannot survive in our naturalistic-scientific system, but this is the very reason why it is important to reclaim it; it is one way through which we can challenge the grounds upon which digital enslavement has been permitted and enabled to flourish. Rather than being fundamentally unfair, as he argues, it is instead very much because our senses of responsibility have indeed been shaped by the historical evolution of our social and cultural systems, that they are able to provide a position from which to judge the rights and wrongs of the creation, sale, and use of digital tech; it is our moral responsibility that warns us of our cyborg present/future.

This concluding chapter draws on three main dimensions of moral responsibilities in the context of digital tech: to ourselves, to others, and for what? It then applies them to both individuals and to communal/collective entities (Walzer, 1983; Williams, 2006). If we are not to some extent responsible to ourselves as individuals, we can never flourish and be of benefit to others; it is our own responsibility to serve as examples and use digital tech safely and wisely. It is also our own responsibility to ensure that what we do is consonant with our moral values and principles. This requires us each to consider what these values are. We begin to act ethically by thinking about the importance of ethics, and what our responsibilities should be, based on that reflection. Moreover, it is these things that we need to get right before we engage in seemingly endless discussions that reinvent the wheel in global summits, or seek to implement new interventions on the ground that claim to be in the interests of the most marginalised.

Responsibilities to ourselves

How we choose to interact with digital tech as individuals can have profound influences over others, both through our formal employment (choosing to work for large digital companies, or instead for small civil society organisations), and through our personal and social activities (volunteering for a helpline supporting the survivors of online sexual harassment, or instead spending our time trolling others and enjoying online pornography). We can choose to focus on gaining as much wealth and power as we can through digital tech, or we can choose a mendicant lifestyle living off the alms of others. However, such choices are not available to the poorest and most marginalised, especially in a world that exploits their labour, both in the physical world and online. Our main responsibility to ourselves ultimately remains to our own health and survival, and recognising this has huge implications for our interactions with digital tech. Do we use it wisely in our own best interests, or do we use it, consciously, or unintentionally, to harm ourselves, and others in so doing?

Likewise, organisations and communities of all kinds have internal responsibilities to themselves and their staff or members with respect to the use of digital tech. Companies have responsibilities to their shareholders and employees; UN agencies are responsible for their actions to their members, including states and other entities; ethnic and caste groups are responsible for supporting and maintaining the lives of their members. It is thus wise in many circumstances to have formal guidance on our organisational responsibilities in using digital tech, to ensure that these technologies are indeed used safely and securely by all.

Responsibilities to others

Traditionally across most societies there has tended to be a gradation in the intensity of responsibilities from the nearest and dearest to spatially distant others. Our most important individual responsibilities are thus often seen as being to our children and parents, then to work colleagues and neighbours living nearby, and then to those at much further distances from us. In the midst of our 'rich neighbours' it is all too easy to forget the 'distant poor'. However, over the last half-century as digital technologies have very rapidly been used to collapse the contexts of space and time, this construct of socio-spatially determined responsibilities has shifted dramatically (Kitchin, 2023). Our closest others increasingly appear to be our 'friends' on social media, or those who inundate us with emails and texts expecting immediate responses, often at the expense of traditional 'family responsibilities'. Increasingly we engage with others, now digitally nearby but once at a physical distance. We nevertheless remain responsible to others for our actions in designing or using digital tech, and we can choose to do so within a wide spectrum from wonderfully good to deeply evil.

Likewise, organisations and communities also have responsibilities to others. This applies at all scales, from small civil society entities to large global corporations, nation states, and international organisations. Thus, although UN agencies are primarily responsible to their members (primarily states, but for some such as the ITU also companies), they also have responsibilities to other UN agencies and indeed the individuals and communities with which they work. UN Women thus has responsibilities to the collectivity of women, UNHCR to refugees, and UNICEF to children. Companies have primary responsibilities beyond themselves to their suppliers and customers, but increasingly they have also accumulated wider responsibilities to the societies and environments where they operate. Modern corporate social responsibility (CSR) emerged in the 1950s (Bowen, 1953; Cragg *et al*., 2009), and its rapid evolution has itself in part been enabled through the use of digital tech. There are, nevertheless, differing views about the value of CSR and corporate social environmental responsibility (CSER), from those who see it as being embedded deeply within their organisational cultures to the other extreme where it has become primarily a means of boosting their brands and increasing their profit levels (see Chapter 2).

Responsibilities for . . .

The question of what someone or some collectivity has responsibility for is often considered in terms of historical, contemporary, and future responsibilities.

Attention in philosophy has particularly been paid to retrospective responsibility, focusing on moral responsibility of actions in the past; whether we did what we ought to have done, linked to the punishments that should be meted out for wrongdoing (Cowley, 2014). This is fraught with difficulty when a society's values and expectations change rapidly over time. The most important response is that we should learn from the past, and ensure that we do not make the same mistakes in the present that we now condemn our predecessors for making.

In contrast, prospective responsibility focuses on what we do in the present with implications for the future, and indeed our potential liability in the future past. This can be with respect to our own moral values, the responsibilities attached to the roles we have, or what is expected of us as human beings within the societies in which we live. As Duff (1998, unpaginated) observes:

> To say that I am responsible for some foreseen effect, or for a harm which I did not prevent, is to say that I should have attended to that effect or to that harm in deciding how to act; our retrospective responsibilities are partly determined by our prospective responsibilities.

This is exactly the type of responsibility that applies in the context of scientists' creation of digital technologies without sufficiently considering the potential harms of their inventions (see Chapter 1). It is thus essential that as many possible future effects of a digital intervention as is feasible are considered before work is done on creating it. It is difficult to reach any other conclusion than that those who are designing the latest generation of digital technologies, including AI-controlled robots and cyborgs, are entirely responsible for the harms that they will cause. They need not have designed them as they have.

There are many ways of considering what different types of organisation might be considered responsible for. Within the corporate private sector quite clear guidance exists on good practices. Fernando (2024), for example, suggests that corporate social responsibility (CSR) is a self-regulating business model addressing four main elements: finances (the viability of a company), ethics (acting fairly and ethically), environmentally (environmental sustainability), and in terms of philanthropy (contributing to society) (see also Clarke, 2017). Within the UN, it is rather more difficult to define core responsibilities. Indeed, the words responsibility or responsibilities only occur 13 times in its foundational Charter, and in most of those cases its meaning is very generalised. However, as noted at the beginning of Chapter 3, signatory governments agreed in its opening paragraph that its core focus, and thus in a sense what it is responsible for, should be preventing the scourge of war, reaffirming faith in fundamental human rights, establishing conditions where international justice can be maintained, and promoting social progress and better standards of life. There is a long history of philosophical discourse on the purpose and responsibilities of government, but, in essence, these are generally seen to be to protect, both externally and internally, and to provide goods and services that cannot readily be supplied by individuals and the private sector. Different forms of government offer very divergent ways through which these two basic responsibilities can be delivered (Zwolinski, 2014). As the extent of civil society has increased over the last century, so too have notions about what it should be responsible for, in part as the result of the abrogation of government responsibilities required by neoliberalism. In general, civil society's main responsibilities are usually seen as being around advocacy (including campaigning and citizenship), service provision (especially for those beyond the reach of governments and the private sector), and accountability (holding governments to account) (Cooper, 2018).

The following two sections develop this broad framework in the context of the individual and then the multi-sector responsibilities that need to be addressed if the most marginalised are to benefit from using digital tech. These apply across most contexts, from discussions in global summits, to delivering small-scale interventions on the ground. Each section provides summaries of the main areas of responsibility to be considered alongside illustrative examples of practical actions that can be taken to enable the poorest and most marginalised to benefit. These are relevant across all areas of our individual and collective action, whether in implementing interventions on the ground or enjoying the policy-making processes in reaching international agreements.

Our individual responsibilities to ourselves and to others

If we cannot free ourselves from the slavery of digital tech, we will never be able to help others free themselves from the same tyranny. This section outlines some of the most important areas of responsibility that individuals therefore need to address in both their personal and professional lives to begin to mitigate the harms of using digital tech so that their undoubted benefits can be achieved. Digital tech can indeed be created and used beneficially by the poorest and most marginalised; we all need to become enlightened and act wisely to help them achieve this.

Digital tech in our personal lives

We are responsible to ourselves for maintaining our health and acting wisely in accordance with our chosen aspirations and ethical outlooks. It is essential to begin by remembering that we can live perfectly fulfilled and indeed exuberant lives without using any digital technologies at all. For most people, the real is, and will remain, better than the virtual, although for some the cyborg blurring of realities is a welcome enhancement to their daily lives of drudgery. Paraphrasing Marx (1844), digital tech has become the opium of the people. To escape the chains of digital slavery, or the kiss and betrayal of digital tech, we need to take conscious actions. If we allow digital tech, and AI in particular, to serve as our digital memory, and cease using our brains to process and remember things, are we not in danger of creating a society where the majority of people become beset by early onset digital dementia that ultimately makes it completely unable for them to function (Spitzer, 2014)?

The following ten principles highlight some of the most important areas of responsibility that can begin to help us reclaim our humanity and escape the control of the Digital Barons, especially if we share them with our nearest and dearest. Our individual actions, example, and advocacy can have a direct impact on the producers of the technologies we use, and can help shape whether they are disempowering and enslaving, or whether they can indeed be crafted in the interests of all, and especially of the poorest and most marginalised.

1. *Being good examples to others.* There is no good espousing an end to digital enslavement while continuing to be addicted to its harms. If we really believe in serving the poorest and most marginalised through digital tech, we have a responsibility to walk the talk.
2. *Living in and protecting Nature.* It is crucial to recall the harms that our everyday use of digital tech causes to the lithosphere, hydrosphere, biosphere, and atmosphere. Many of these affect the poorest far more than they do the richest. Enjoying our direct interactions with Nature through all our senses is crucial to remaining human, and costs little.
3. *Consciously considering the costs and benefits of using every piece of digital tech we own.* It is often hard to think that we do not actually need to be on social media. The single thing we can most easily do to reduce the control and wealth of the Digital Barons is simply to stop using their apps and services ourselves.
4. *Our use of language.* Using terms such as 'bridging the digital divide' (see Chapter 2), 'frontier tech', and suggesting that technology actually causes anything or will deliver the Sustainable Development Goals (SDGs), not only provides an insight into how we think, but it also hides the truth about the creation and use of digital tech (Unwin, 2022a). Our language needs to reflect our interests in working with and in the interests of the poorest and most marginalised.
5. *Repairing rather than replacing.* One of the most important ways we can reduce our environmental impact is by insisting on the repair of our digital devices rather than simply replacing them (see vignette by Vallauri, p. 164). However, always switching devices off, minimising our use of high-definition video, not feeding the beast of AI, and reducing all our digital waste are also among the many other important actions we can take to reduce our environmental impact.
6. *Retaining real rather than virtual transactions.* If we do not insist on using cash, we will lose it, and this loss will have the greatest impact on those who are excluded from having bank accounts or access to online money systems. Moreover, digital transactions are one of the most valuable ways through which tech companies and governments extract and combine data about us. Limiting such transactions limits their power over us.
7. *Using digital tech safely, wisely, and securely.* As digital tech systems become more complex and powerful, users become much more susceptible to online harms, from trolling and surveillance, to phishing attacks, fake videos, financial scams, revenge porn, and ransomware attacks. The most marginalised are often the most vulnerable since they rarely know how best to protect themselves.

8. *Resisting the imposition of having to use certain digital tools.* One of the most insidious and enslaving characteristics of digital tech is the way in which it is increasingly used to pass the physicality of labour down to the end user or customer. Frequently, it is now essential for the purchaser or service user to download an app and enter information about themselves to gain a service from governments or companies. We should instead insist on using our own existing digital systems and software and require powerful others to accede to our specifications rather than them imposing their systems on us.
9. *Using Open tools and resources where possible.* Advantages of using Open software, data, and resources can include lower cost, interoperability, enhanced security, collective support, transparency, and independence, but at the same time the ways through which private sector companies often exploit Open data and systems in their own interests need to be recognised (see Chapter 6).
10. *Protecting multiple identities.* As individual humans, we have multiple identities, depending in part on the contexts in which we find ourselves at any space-time conjuncture. Yet, the use of single digital identities is constantly being forced on us, as with Aadhaar in India (see vignette by Al Dahdah, p. 121), and it is becoming much more difficult to retain separate, disconnected identities for ourselves. The perceived need for single digital identities, controlled by authorities as part of their apparatus of surveillance, further marginalises those who wish to retain the multiple identities with which they were born or have created.

The right to repair

Ugo Vallauri

(Co-Director, The Restart Project and Co-Founder, Right to Repair Europe)

A man in an informal settlement in Nairobi sells a rusty, beat-up, used yet useful power extension cord. In central London someone drops a still functioning laptop into a recycling skip, smashing the device so that no one will be able to extract data from it – while preventing it from ever being used again.

E-waste is one of the fastest-growing streams of waste globally; 62m tonnes were produced in 2022, with only 22% recycled. However, most environmental impact of the devices powering our always-on lives – laptops, tablets, smartphones – occurs before we've ever switched them on, during mining and manufacturing. Yet, in the most affluent societies we've been discouraged from repairing and reusing, to fuel the throwaway economy of convenience and same day deliveries. Even 'circular' economy projects tend to focus on recycling so that overproduction and overconsumption can continue.

For the last 15 years, the community repair movement has shown that there's another way: that it's possible to give devices a second lease of life, to create abundance and digital access out of Big Tech's deliberate attempts to make products obsolete, with extortionate pricing of spares, repair-averse design and increasingly using software as a weapon.

People, not electronics, are at the heart of Restart Parties and Repair Cafes: organisers, volunteers, participants, united by the curiosity to learn collaboratively how everyday objects work and how to find alternatives: workarounds, compatible spare parts, Free and Open Source operating systems to counter the constant upgrade cycle. We fight for the universal right to repair – meaning that everyone should have the right to access affordable repair options. The movement is increasingly respected by policymakers, and laws are slowly starting to change: from design requirements for future smartphones, to user-replaceable batteries in consumer hardware, to consumers' right to repairs at 'reasonable' prices.

Technology and hype cycles are, though, much faster than policy changes, so while in Europe we're busy discussing 'horizontal' repairability requirements to improve ecodesign of all products, manufacturers are investing heavily in trying to convince consumers that they must upgrade for new AI features. Elsewhere, in Africa, Asia and South America repair remains vital: a necessity, not a privilege.

In our professional lives

For some people there is little distinction between their personal and professional responsibilities, and it is also not easy to generalise around very diverse types of employment and occupations. However, drawing especially on the material in Chapters 4 and 5, a further ten responsibility principles relating to our professional lives are important in shifting the balance away from digital enslavement and towards a system that does indeed focus on serving the needs and interests of the world's poorest and most marginalised. In essence, we are responsible to all those we encounter in the places where we work, including both our colleagues and those clients or customers for whom our labour power is used. Should our responsibilities not be to do them good, in whatever ways we or our employers define this? We are also responsible for sharing our ethical outlooks that in part determine our advocacy and actions.

1. *We have choices over those we work for.* The starkest choice to be made is over the sort of job we choose to do, and for whom we wish to work. In essence, we can work in and for the big tech companies, or for poor people and marginalised communities. The former can bring rich financial rewards for 'me'; the rewards of the latter, focusing on the 'we', are much more intangible.
2. *Taking responsibility for the harms caused through digital innovations.* This book has throughout reiterated that digital tech by itself causes nothing. It is the people who design, build, and sell it who are the primary cause of change. It is their outlooks and actions that need to be changed. All too often, these people are delighted to take the credit for successes, but blame others for their failures and the harms that have been caused through poor implementation. It is time for scientism to be rejected once and for all with respect to innovation

in the digital tech sector. Thorough and wide-ranging potential impact analyses that rigorously examine the many possible harms of any major new tech development must be undertaken especially when working with the poorest and most marginalised (see, for example, MRC, 2012; NHS, 2018; UNESCO, 2022c; n.d.; Floridi, 2023; UNODET, n.d.).

3. *Ensuring we are well informed.* Many people working in digital tech for development have surprisingly little knowledge about previous initiatives undertaken in the field. This is inexcusable, particularly in the academic sphere where tracing the origins of ideas has long been seen as the hallmark of scholarship (Sein *et al.*, 2019; Schelenz and Pawelec, 2022; Masiero, 2023; Thapa, 2025). Vast sums are spent on rolling out digital tech initiatives in economically poor countries, where past experiences of similar interventions show clearly that they will never fulfil their publicly stated ambitions and serve the interests of the poorest and most marginalised (The Digital School, 2023; 2025). Knowing what has gone before is a fundamental prerequisite for any organisations claiming to deliver universal digital solutions that will benefit everyone (see the following examples: Weiler *et al.*, 2005; ITU, 2011; Sadowsky, 2012; Nokia, n.d.; Intel, n.d.). It is irresponsible to make the mistakes that others have made in the past, and to be unaware of many of the initiatives that have gone before (see vignette by Hollow, p. 126).
4. *Being responsible to the poorest and most marginalised.* The necessity to work with, not for or on, others is another recurring theme within this book, and is absolutely central to effecting beneficial implementations of digital tech in marginalised contexts (see vignettes by Harindranath, p. 96; Longmore, p. 27; Vallauri, p. 164; and Juliette Unwin, p. 85). The only way that such initiatives can become relevant and sustainable is to work hard at consciously and humbly reducing the 'me' and paying more attention to the 'we' in delivering digital interventions.
5. *Minimising our own organisational use of digital tech.* If digital tech in general is used to drive inequalities, then it makes sense to minimise our own organisational use of digital tech, and to ensure that staff can also have sufficient freedom to engage in the world of Nature without digital tech. Moreover, far too often organisations insist on a particular digital environment that all staff, and frequently those others who also engage with the organisation, have to work in, rather than creating systems with which people can interact on their own terms and in their own ways.
6. *Limiting the wastage of digital tech use.* One of the most frustrating issues around digital tech within the workplace is the way through which its use often dramatically increases wastage, and paradoxically restricts the time that most staff have to do important work. This is typified by the plethora of unwanted e-mails, social media and group messages, and inefficient online video meetings. The poor and marginalised cannot afford such wastage.
7. *Promoting the right to be unconnected.* Another recurring theme of this book is that the right to be unconnected is at least as important as any putative right to be connected. If the connected world is one that is invisible to a blind user,

or unintelligible to someone who cannot speak its languages, or is used to promote cultural practices dissonant with those of an ethnic minority, then it further marginalises the already marginalised.

8. *Questioning who benefits from our innovation.* If we are seeking to be responsible for helping the world's poorest and most marginalised improve their lives, should it not also be our responsibility to focus primarily on using tried and tested technologies and ensuring that these are indeed made available for such groups to use (Chapter 5)? In some circumstances, further technological innovation may well be needed to achieve this, but it is important to resist the neophiliac rush to innovate just for the sake of innovation. Above all, no-one should innovate or experiment 'on' people in poor and marginalised communities. We should never take pride in 'building the plane while flying it'.
9. *Ensuring universal access and usability.* If the digital technologies we craft are not equally available to all, they will further increase inequalities. As far as possible, we should therefore seek to ensure that the poorest and most marginalised should be able to use them without any additional cost or expenditure. Great strides have indeed been made in this direction over the last 20 years, particularly for people with disabilities, but much more remains to be done (Eid, 2017; Awde *et al.*, 2022; see also Gardezi vignette, p. 34).
10. *Thinking (and reading) before acting (and sending messages).* A final principle of responsibility in our professional digital lives is that we should always speak less than we listen, than we think, and than we read. Far too few people working in the field of digital tech for development have the sufficiently deep understanding of the critical interface between theory and practice (Habermas, 1974) that is essential for wise policy making and action (see vignette about Nigel Hickson, by Hellerstein *et al.*, p. 177).

Multi-sector responsibilities

The private sector, governments and civil society need to work effectively and collaboratively together if the poorest and most marginalised are indeed to benefit from the use of digital tech. All must be together at the table. This is one of the core responsibilities that entities in each sector have, and there remain far too few platforms or programmes that do this well at a global, regional, or national level. All too often companies and governments dominate the global policy landscape (see Chapter 3), without sufficient attention being paid to the leavening voice of civil society, despite advocates of the IGF claiming that this has indeed provided some basis for effective 'multistakeholder' discussions. Where 'responsibilities' have been referred to in most recent international agreements concerning digital tech, these are rarely articulated clearly, and the different responsibilities of each sector remain ill-defined (see Chapter 3). All too often, global partnerships for the use of digital tech in development are created without any clear structure or understanding of the fundamentally different roles and responsibilities of the types of partner involved (Unwin and Wong, 2012). What these roles are depends in part on our ethical outlooks and in part on our political philosophies (O'Neill, 1990;

Rawls, 2001; Zwolinski, 2014). However, regardless of the precise philosophical positions that we hold, it is possible to identify six broad areas of responsibility for each sector that, when combined, can optimise the likelihood that digital tech interventions may indeed be developed wisely in the interests of the poorest and most marginalised. Different political systems will prioritise the importance of each of these in varying ways, and the increasing divergence of political ideologies over the last 25 years has meant that there is now much less international agreement around the neoliberal democratic ideologies that underlay the Millennium Development Goals (MDGs) at the start of this century. Overall, this section argues that if the poorest and most marginalised are indeed to benefit from the appropriate use of digital tech, then governments need to accept the responsibility of playing a much more active role in ensuring that this happens, aided especially by the advocacy role that civil society can play in ensuring that justice for all can be achieved. This represents, on the one hand, a rowing back from the excesses of neoliberalism in the USA and many other parts of the world over the last quarter-century, and, on the other, an equivalent shift to greater individual freedoms in authoritarian and dictatorial states elsewhere (Economist Intelligence Unit, 2025). Identifying common ground between these two political extremes will be difficult, but it is not impossible if sufficient people of good will on both sides commit to working together to reach agreements and create structures that reduce inequalities and prevent the dystopic scenarios described in Chapter 6. As O'Neill (2016) reminds us in this chapter's epigraph, we have an important choice to make between action and inertia.

The private sector

The private sector's fundamental role in most societies is to produce goods and services by employing labourers who are notionally free to decide for whom they choose to work. In the broadest sense, companies have responsibilities to their owners and shareholders, their employees (or labour), their consumers (those who buy the digital tech they produce), and the governments of the states where they operate (in the form of taxation). Given these responsibilities, it should be self-evident that it is not necessarily their role to deliver digital technologies to the poorest and most marginalised, especially if they are unable to do so profitably. There should be no expectation that they should do so (although see vignette by Hughes, p. 115).

- *Producing goods and services.* The private sector is the main driver of economic growth, but the arguments in Chapter 2 concerning alternative no-/low-growth and more redistributive models of the economy suggest that this should not necessarily always be their only priority. If the poorest and most marginalised are indeed to benefit, it is the responsibility of digital tech companies to employ enlightened staff (Seth, 2022), develop and produce technologies that are as far as possible universally accessible and repairable, understand how technologies can benefit the world's most marginalised, and then produce them at a price that the poor can afford while minimising negative impact on the environment. This requires a fundamental conceptual and practical shift from most current business practices in the sector (see vignette by Hughes, p. 115).

- *Research and development (innovation).* Companies and universities undertake most of the research and development behind the creation of new digital technologies. They therefore have very significant responsibilities for doing good or harm to those who purchase or use their products. Although most bottom of the pyramid approaches to digital tech have failed, companies should nevertheless be encouraged to focus on research and development that can bring as much universal benefit as possible, while also minimising harms to the most marginalised and to the environment.
- *Wealth generation.* Companies are widely seen as being the main source of wealth generation, although this currently accumulates mainly in the countries where they are headquartered. They need to take greater responsibility for investing in the countries from which their profit is originally generated. Moreover, if digital tech companies wish to walk the talk of 'digital for all', they need to focus much more on implementing solutions with and for the poor and marginalised, and thus seek to place a reduction of inequalities at least on a par with wealth creation in their priorities.
- *Employment (extraction of labour power) and provision of work.* There is widespread disagreement on the potential future impact of AI in the workplace (Mayer *et al.*, 2025), but in the economically richer countries of the world the expansion of employment in the digital tech sector over the last half-century seems to have more than compensated for the reduction in low-skilled manufacturing jobs. Its most important impact has been to increase the productivity of most workers, and digital tech has therefore enabled capital to extract an even greater surplus from their labour force. This has nevertheless varied spatially, with countries that have existing high skills levels tending to benefit most (World Economic Forum and Capgemini, 2024; IMD 2025). Significantly, countries in Africa and South America are generally the least technologically advanced and have the lowest levels of employment in the tech sector (World Bank, 2024e; World Population Review, 2025).
- *Corporate Social and Environmental Responsibilities (CSER).* Chapter 4 highlighted how projects supported by CSR interventions from the private sector have generally made little fundamental and systemic improvement to the lives of the world's poorest and most marginalised, although in some specific instances they have changed the lives of many individuals. They tend to be a palliative rather than a fundamental cure for digital tech-driven inequalities. Nevertheless, a few companies and foundations with a deeply embedded ethical agenda have shown how digital tech might be used to make a fairer world. Digital tech companies should learn from these examples, and should embed these principles across their whole business fabric so that interventions are more sustainable and affordable, both financially and environmentally.
- *Distribution of goods and services.* One of the greatest drivers of inequality in the digital tech sector is the cost of providing infrastructure to the poor

and marginalised. Difficulties faced by fixed and mobile broadband providers in generating sufficient revenues from marginalised contexts have been at the root of many discussions surrounding universal connectivity, especially in contexts where Over-The-Top (OTT) service provision has proven to be most profitable. The privatisation of telecommunication operators in the 1980s and 1990s has not ensured uniform quality of services and has relatively disadvantaged the most marginalised. A strong case can be made to bring this basic infrastructure back into public ownership, and then let the private sector compete over the services delivered through it. Privatisation only makes economic sense if the surpluses paid to management and shareholders are less than the inefficiencies lost through what is often described as inept government management.

Governments and international organisations

It is the fundamental responsibility of governments to serve the interests of all their citizens. Whereas the private sector cannot be expected to deliver to the poorest and most marginalised, it is the role and responsibility of government where necessary to provide such services, to ensure that the most marginalised are not further disadvantaged, to minimise harms of all kinds, and to redistribute wealth where necessary to ensure equality of opportunity and to provide an acceptable standard of living for all. International agencies and the UN system should likewise play a key role in supporting this, where necessary through international agreements and the provision of guidance on good practices to those states that cannot afford the research and development costs associated with innovation. UN agencies, though, are responsible for ensuring that recommended good practices are in the interests of the poorest and most marginalised, and not just for the rich and powerful. The failure to secure a fairer, less harmful and more equitable use of digital tech is not primarily a failure of the private sector. It is mainly a failure of governments and the UN system.

The following are the most important actions that governments and the UN need to take if the poorest and most marginalised are to benefit from their digital tech intervention:

1. *Public services*. The most important role of governments, and indeed the UN system more generally, is to provide peace and security, but also basic human needs such as education, health, and sustenance for all. However, as digital tech is increasingly used to deliver these services, especially through critical infrastructures, governments also have clear responsibilities for ensuring the security and safe use of these technologies. Cyber security, globally, nationally and individually is thus of immense importance, and all too often, especially among the poorest and most marginalised, it receives insufficient attention (Cornish, 2021). There needs to be widespread global recognition that digital tech can be used to do as much harm as it does good, and that the harms require mitigation if the good is to prevail.

2. *International agreements.* It is also crucially important that sound global agreements around digital tech policy and security are forged and enforced, especially with respect to the commons, such as outer space and the deep seas. A strengthened ITU, in the context of an effective CEB for inter-agency co-ordination, seems best placed to take the lead in this area, as it has done for many years with international agreements over the radio spectrum. There is no point in further wastage by the UN Secretariat in creating a parallel system in its Office for Digital and Emerging Technologies, created in 2025, which seeks to take forward the Global Digital Compact, AI governance, and common frameworks over digital public infrastructure (see Chapter 3). Paradoxically, the behaviour of the Trump regime in the USA may provide just such an opportunity for a reconfigured and reinvigorated multilateralism that could be forged by other states seeking to create a better world that is less dominated by US capital and more representative of the needs of the world's poorest and most marginalised.
3. *Regulation.* Regulation in the digital tech sector has for too long tended to serve the interests of the private sector more than it has the world's poorest and most marginalised. This is one of the most important areas where governments, as well as international, and regional organisations can best serve all their citizens responsibly, but, to date, they have consistently failed to do so. Three areas of regulation are of particular importance: (1) provision of infrastructure, (2) reduction of digital harms, and (3) promotion of freedoms in the digital context. If the poorest and most marginalised are really to benefit from using digital tech, then governments at all levels need to shift the balance in their interests, rather than in those of the profit-motivated private sector. This will require much stronger regulatory powers, and a willingness to enforce them. The processes that led to the European Union's Digital Services Act (European Commission, 2022), and its emphasis on seeking to ensure a safe and accountable online environment for all provide valuable lessons for the wider international community on what can indeed be achieved.
4. *Sharing good practices.* There is no such thing as universal best practice, except in the minds of those who promote the notion in their own interests to sell a product, gain prestige, or acquire power. However, governments and UN agencies have a very important responsibility to share evidence and recommendations on good policies and practices relating to the use of digital technologies for all, with an emphasis on ensuring equity and effectiveness. Again, UN agencies have a vital role to play in sharing and developing such good guidance, especially in association with experienced and altruistic academics and practitioners. Minimising the wasteful duplication of effort, and working with governments and like-minded untied donors or foundations to implement such structurally transformative programmes could go a long way to shift the balance away from the interests of global capital towards the world's least advantaged.
5. *Redistribution of wealth.* National governments have the responsibility for ensuring that effective taxation regimes are in place to enable all citizens to live fulfilled lives. However, it is very hard to see any justice in the very

considerable profits accrued by the Digital Barons and those who emulate them, and higher levels of taxation on the digital tech sector in its broadest sense could therefore contribute to the redistribution of wealth in the interests of the poorest. Significant positive benefits could be accrued if such taxation were to be used wisely to deliver interventions focusing explicitly on supporting the poorest and most marginalised. Universal service/access funds provide one mechanism for delivering such support, but it is important to learn from the disappointing early experiences of many such funds (Ladcomm, 2013). Another dimension of responsible redistribution that must be considered is the amount of tax paid by multinational companies to the governments of jurisdictions where they operate (OECD, 2025) but existing regulations do not go far enough to make much impact on redistributive justice in the interest of the most marginalised.

6. *Public utilities in the interest of all.* A strong case can also be made that governments should intervene directly if they cannot ensure that connectivity for all is guaranteed by the private sector, although experiences with the roll-out of Australia's controversial National Broadband Network suggest that great care is necessary in such interventions (Walker, 2024). For small island states with insufficient demand to enable competition to drive down consumer prices, it also makes sense for government monopolies to deliver connectivity, although these too require careful management and transparency. Such interventions will be expensive, but if digital connectivity is seen as being an essential utility for all, then governments need to step up to the mark to ensure such delivery.

In summary, governments have the responsibility for focusing explicitly on serving all of their citizens through digital tech if the poorest are indeed to benefit. For too long the digital tech sector has been controlled primarily by private sector interests that have driven substantial global inequalities, and enabled the Digital Barons increasingly to enslave us all. Anyone serious about ensuring that the poorest and most marginalised can benefit from using digital tech needs to recognise that the time has come to replace neoliberalism with a new model of political economy in which governments take greater responsibility in ensuring that priority is placed on equity rather than growth (Xi, 2014–2025).

Civil society

The above sub-section on the need for wise government in the interests of all presumes that those with political power do indeed use that power wisely with respect to digital tech. This does not always happen, although in my experience it has always been possible to find people of good intent to work with in any government or company. More formally, one of the prime roles of civil society in the multi-sector responsibilities model summarised here is indeed to hold governments to account. As with the private and public sectors above, six broad areas of responsibility and indicative actions are particularly pertinent for civil society organisations committed to delivering beneficial digital interventions for the poorest and most marginalised.

1. *Holding governments and international agencies to account.* The most important role for civil society is to promote the voices of citizens in their collective dealings with government. At their best, they can hold governments to account (Locke, 1988), and in democracies have an important role to play in supporting citizens' voices through elections. However, in much international discourse on digital tech, as outlined especially in Chapter 3, the role of civil society has not yet been sufficiently powerful to achieve this in the face of the economically and politically dominant influence of the private sector. The voice of civil society needs to become very much stronger, not only in holding national governments to account, but also in speaking truth to power at the global level with regards to the impact of digital technology on the lives of those who are increasingly enslaved through its use.
2. *Advocacy.* Closely aligned with the previous point, is the important responsibility that civil society has in advocating with and for the poorest and most marginalised. As noted previously, entities such as the APC, the Just Net Coalition and Diplo have all played very important collective roles in international discussions on digital tech and especially on Internet governance, but this needs to be considerably amplified. Moreover, different civil society organisations are often quite divided in their views, and it remains a challenge to identify common goals and agendas around which diverse civil society organisation can unite.
3. *Serving the poorest and most marginalised.* Many civil society organisations, and especially small NGOs, have historically taken upon themselves the responsibility of delivering goods and services on the ground to those falling beyond the ability and willingness of governments and the private sector to provide them with such resources. There are countless small civil society organisations across the world promoting the use of digital tech for health, education, skills development, and numerous other purposes, which have had a direct beneficial impact on the lives of individuals. However, there is a fundamental tension in this model because the existence of such organisations can hide the responsibility of governments to make provision for these actions themselves, and in very few cases does the work of these organisations effect the structural changes that are so necessary if the poorest and most marginalised are really to benefit.
4. *Championing diversity and youth engagement.* The diversity of civil society organisations reflects the diversity of marginalisation, especially around themes of gender, disability, ethnicity, political status, age, and race. In many countries coordinating such responsibilities can be a real challenge, with different civil society groups competing against each other for limited resources, and a strong case can be made for enhancing the capacity of such organisations to work together collectively to effect positive change. As the vignettes by Longmore (p. 27) and Morenets (p. 55) illustrate, youth engagement is not only of central importance to decision-making around the future use of digital tech, but it can

also serve as a wider model for effective consultation and consensus building as noted by Edwards in Figure 7.2.

5. *Transparency and openness.* Civil society has long championed the importance of communal solutions, the value of Open digital policies and practices, and the need for transparency across government and the private sector. Chapter 6 has nevertheless highlighted how the private sector has often subverted Openness to serve its own interests, particularly with respect to data acquisition, and despite the intuitive value of Open content and software, it has proved difficult to reach global agreements that have in practice really benefitted the poorest and most marginalised. Civil society must nevertheless retain and nurture this responsibility in the interests of the many rather than the few.
6. *Freedom.* Ultimately it is civil society that has the responsibility for maintaining and cherishing the values of freedom that abolitionists proclaimed so vehemently in the nineteenth century. The continuing extent of modern slavery, for example, is testament that there is much work still to be done to eliminate human trafficking and enforced labour. However, as our minds and increasingly our bodies become progressively more unfree, subjected to control and exploitation by the Digital Barons and those who aspire to replace them, it is the responsibility of civil society not only to speak out strongly in defence of digital freedom, but also to campaign ever more vigorously in support of the world's poorest and most marginalised. Paradoxically, it may well be the unconnected, the neurodiverse, those with disabilities, and ethnic minorities, those still beyond the reach of those who wish digitally to enslave them, who retain the secrets of how we can escape from the underground and reclaim our harmony with Nature.

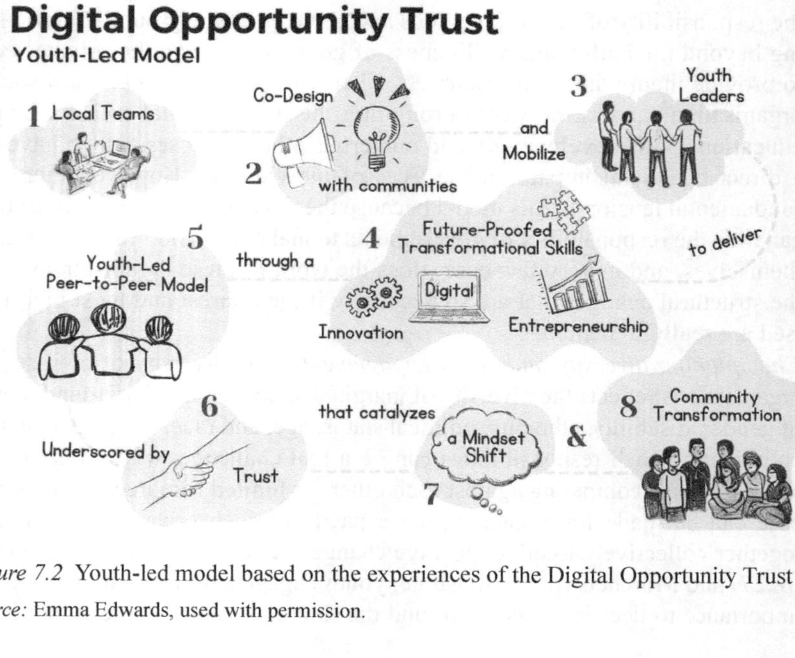

Figure 7.2 Youth-led model based on the experiences of the Digital Opportunity Trust

Source: Emma Edwards, used with permission.

Priorities for change

Finally, there are six practical changes in our thinking, our use of language, and in our practical interventions that can begin to shift the balance away from the dominant economic growth agenda and towards a more equitable approach that will enable the poorest and most marginalised to benefit appropriately from the use of digital tech. Digital technologies can undoubtedly be used for good, in whatever way we choose to define it, but for this to be achieved the significant harms associated with such use must be mitigated.

First, as so many of the vignettes in this book have argued, emancipation begins with us. *Those of us working in digital development need humbly to become the servants of the poorest and most marginalised.* We constantly need to be reminded that we should not just be working 'for', or worse still 'on' them, but rather 'with' them. It should not primarily be about 'me', but rather about what 'we' can do together (see vignettes by Harindranath, p. 96, Longmore, p. 27, and Juliette Unwin, p. 85). Our use of language matters.

Second, we must always recognise that *technologies are designed, created, sold, and used to serve particular sets of interests*. There is nothing 'natural' or 'neutral' about them. The notions of technological determinism and digital instrumentalism must be rejected and challenged at every opportunity if the poorest and most marginalised are to break the shackles of their digital enslavement. They cannot fully benefit unless the technologies are designed with their interests uppermost in mind. Scientists are as responsible for the harms of the technologies they bring into being as they are for any benefits, and we need constantly to articulate the harms if any good is to arise. As Tarmizi's vignette (p. 175) emphasises, our intervention should not be driven primarily by interests in the technology, but rather 'Let's begin by agreeing on the societies we want, then apply the appropriate technologies to make them thrive.'

It's about what technology can do for society

Mohamed Sharil Tarmizi

(A curious soul, constantly experimenting and trying new things)

Over two decades ago, when I was with the Malaysian Communications and Multimedia Commission (MCMC), I wrote an article in a UN publication entitled Digital Reach (ITU, 2005). It was called, 'Malaysia Bridges the Digital Divide' – a story of how Malaysia set up basic Internet access for rural communities. At the time, the world saw information and communications technology (ICT) expanding into the concept of the Information Society. We saw two World Summits on the Information Society (WSIS) in Geneva (2003) and Tunis (2005). The Millennium Development Goals (MDGs) had also just been adopted by the UN.

The Malaysian story was about how the '*kedai.com*' ('shop.com') would bridge the digital divide, as we then saw it, and transform people's lives, particularly in rural areas. At the heart of it, two desktop PCs were placed in convenient areas adjacent to an existing village sundry shop. The PCs would have a shared Internet connection of 128kbps. It was a bold move then because emails, the Internet, and the World Wide Web only became available in major cities in Malaysia in the mid-1990s.

In 2010, as part of the National Broadband Initiative (Malaysia Government Portal, 2025), I was involved in the initiative to provide netbooks and WiFi connectivity to nearly 1.7 million high school students. It was part of Malaysia's desire to 'bridge the digital divide' and improve broadband accessibility. Whilst we still provided community broadband access, we had moved up the chain to provide targeted individual access to ICT (see MCMCTV, 2013, for our videos on the theme of 'Everything is Possible with Broadband'; and Digital News Asia, 2013).

It is now 2026; I left MCMC over a decade ago and Malaysia is still providing free collective broadband access to rural and underserved communities. What has changed? Arguably, a lot. And yet, fundamentally not so much – instead of 128kbps PCs and ICT, it's now 5G and the digital economy (MCMC, 2024). The technology has evolved and yet inequalities still remain. In the end, it's never about the technology but it's about what technology can do for society. Let's begin by agreeing on the societies we want, then apply the appropriate technologies to make them thrive.

Third, it is important that we craft accurate rigorous *understandings of the outcomes of digital interventions*, and share these even more widely than they are at present so that the most marginalised do indeed benefit from the many good practices that currently exist. There should be absolutely no excuse when well-intentioned people, who claim to be developing novel interventions, make the same errors that countless others have made before. Likewise, there should be no excuse when academics who have not read sufficiently widely claim that their research is innovative and ground-breaking, when earlier very similar work remains uncited in their publications (Unwin *et al.*, 2023). Measuring outcomes is not always easy, but the rhetoric around the success of digital interventions should never be based on inputs (devices introduced, cables laid, or financial pledges) or outputs (teachers trained, courses delivered, or lines of code written) alone. Instead, every digital intervention needs to be judged on the real outcomes that it has produced, on how it has improved the lives of the world's poorest and most marginalised people, and whether such outcomes could have been achieved at better value for money using a completely different approach (see vignette by Hollow, p. 126).

Fourth, *we need to reassert the importance of Nature and the physical environment for human existence*. We all construct the idea of Nature in our own ways, influenced by the social and cultural structures within which we live, and our own diverse experiences of the physical environment. Nevertheless, it does appear that practical engagement with and within the physical environment remains of particular importance for human health and well-being. Not only do we need much more rigorous and holistic approaches to measuring the relative harms and benefits of digital tech use on the physical environment itself, but we also need to reassert the importance of spending time digitally disconnected, engaging with the natural world, and rediscovering that the real world is still vastly better than the virtual world.

Fifth, we need to revise the terminology of multistakeholder partnerships to reflect *truly multi-sector interventions, that give equal status to civil society, alongside governments and the private sector*. This will require fundamental restructuring of the failed UN system, an emphasis on sectors rather than generic stakeholders, and rigorous partnership frameworks that make transparent the contributions and expectations that each partner has of their engagement. It applies especially at the level of interventions on the ground, but is also highly pertinent in the international arena where rhetoric and reality are often considerably far apart (see vignette in memory of Nigel Hickson, Hellerstein *et al.*, p. 177).

Nigel Hickson: A digital life well lived for others

Judith Hellerstein, Alejandro Pisanty, Lousewies van der Laan, Pari Esfandiari, Emily Hickson, and Tim Unwin

Figure 7.3 Photo of Nigel Hickson

Nigel Hickson was writing a case study for this book until his untimely death prevented its completion. He had a distinguished career for the UK Government and within ICANN. Above all, he devoted himself generously to others, encouraging us all to strive so that everyone can benefit from the Internet through effective multi-sector partnerships. These reflections from some of those who worked with him and counted him as a friend, are shared here as a model for others to follow. The photo shows Nigel enjoying himself after a long day of discussions at the ICANN meeting in California in 2014.

> Nigel had the kindest heart and was always willing to devote time and energy to help others. He brought light into the world, and I will miss his humour and his ability to put a smile on anyone's face. He was always there to impart wisdom and teach ICANN fellows, friends, and colleagues new things, but most of all, he was there for his friends and family.
>
> (Judith Hellerstein)

> Nigel Hickson was a remarkable man. For me, his most salient feature was how he managed to be of one piece. Whatever organisation he represented, whether the government or another entity, his principles and broad view meant he could state a position and fight for it without regret or contradiction. He was a good man by that most critical count.
>
> (Alejandro Pisanty)

> Politicians and policymakers have a hard time understanding that something as global, vital and complex as the Internet could be run by an open inclusive multi-stakeholder community. Nigel was one of those exceptional people who could build a bridge between those worlds and make newcomers feel not just welcome, but part of this crucial community.
>
> (Lousewies van der Laan)

> Nigel was a rare anchor in a world whirling with digital illusion – grounded, generous, and quietly wise. For years, we met virtually, but one afternoon in London, he stepped out from the screen into the real: no longer a digital image, but vividly human. That day, a deep friendship took root. In a world wired for noise, Nigel listened as if every voice mattered – and in doing so, reminded us it did. He moved without ego, thought deeply, and stayed kind. He wasn't just present – he was true.
>
> (Pari Esfandiari)

> A born adventurer, in his youth Nigel hitchhiked and trekked all over the world, meeting new people and exploring new cultures. The birth of the Internet allowed those adventures to continue, building the physical and virtual community that mourn him so much today. To his mind, digital technologies were overwhelmingly a force for good, with the beneficial potential for humanity correlated with the rigour and values of the people and institutions that governed them.
>
> (Emily Hickson)

If we seek to be emancipated, Nigel's life well spent in the digital service of others is an example we would all do well to follow.

Finally, we must take urgent steps to *forge our freedom from the enslavement into which we are being seduced by the Digital Barons*, and encourage others to act likewise along the lines suggested in this book. If we do not act urgently, our lives will become ever more surveilled and controlled, our minds ever more dysfunctional, our labour ever more futile, and the physical world that is so vital to our existence ever more despoiled. This applies to everyone, but particularly to the most marginalised, the voiceless, and the powerless. It is our responsibility above all to ensure that they become central to our practices, that their voices are heard, and that they have the power and ability to use digital technologies safely, wisely and securely in their own interests. We cannot afford to keep on making the choice of inertia (O'Neill, 2016), and must act wisely and urgently to reject the mental and physical enslavement that comes from being born from an idea. We must resist developing the taste for a world where parent-less cyborgAIns, without a real human body and blood, encourage us to believe that we are content with our lot. We can start to do so by engaging in, and caring more for, the physical world of nature, by addressing the unacceptable inequalities caused by the design

Figure 7.4 Tarn on southern slopes of the Schwartzhorn looking towards the Eiger in the Berner Oberland, Switzerland

Source: Tim Unwin.

and use of digital tech, and remembering that anyone's death diminishes each one of us.

> Look into it more carefully! Why, we don't even know what living means now, what it is, and what it is called? Leave us alone without books and we shall be lost and in confusion at once. We shall not know what to join on to, what to cling to, what to love and what to hate, what to respect and what to despise. We are oppressed at being [people–people] with a real individual body and blood, we are ashamed of it, we think it a disgrace and try to contrive to be some sort of impossible generalised [person]. We are stillborn, and for generations past have been begotten, not by living fathers, and that suits us better and better. We are developing a taste for it. Soon we shall contrive to be born somehow from an idea. But enough; I don't want to write more from "Underground".
>
> (Dostoyevsky, 2021, 2: X; first published, 1864)

Bibliography

All material was accessed between 1 September 2024 and 1 August 2025. In most instances, only the first-named author (followed by *et al.*) is cited by name when there are more than ten authors for a publication.

4BM (2025) About us, https://4-bm.africa/about/.

Abelson, H., Miyagawa, S., and Yue, D.K.P. (2021) On the 20th anniversary of Open CourseWare: How it began, *MIT Faculty Newsletter*, 33(5), https://fnl.mit.edu/may-june-2021/on-the-20th-anniversary-of-opencourseware-how-it-began/.

Acemoglu, D. and Johnson, S. (2024) *Power and Progress*, New York: PublicAffairs.

Acemoglu, D. and Robinson, J.A. (2013) *Why Nations Fail: The Origins of Power, Prosperity, and Poverty*, New York: Crown Currency (Reprint edition).

Age UK (2023) *Avoiding Scams*, London: Age UK, https://www.ageuk.org.uk/siteassets/documents/information-guides/ageukig05_avoiding_scams_inf.pdf.

Aggarwal, S. (2018) *Freedom to Fail: Lessons from My Quest for Start-up Success*, Noida: HarperCollins.

Agile Alliance (n.d.) What is Agile?, https://www.agilealliance.org/agile101/.

Ahmed, S. (2022) The marginalized and Critical Theory: dialectics of universalism, *International Critical Thought*, 12(2), 305–326, https://www.tandfonline.com/doi/full/10.1080/21598282.2022.2070857.

Ahmed, S. (2025) From innovators to oligarchs: The role of Big Tech in Trump's America, *The American Bazaar*, 25 January, https://americanbazaaronline.com/2025/01/20/big-tech-oligarchs-in-trump-america-458527/.

Aibinu, A. *et al.* (2022) In defense of Fabrizio Hochschild, 26 January, https://www.passblue.com/wp-content/uploads/2022/01/Letter-in-defense-of-Fabrizio-Hochschild_updated-27-Jan.pdf.

Aitkenhead, D. (2025) Melinda French Gates on divorcing Bill and giving away her billions, *Sunday Times Magazine*, 13 April, 11–15, https://www.thetimes.com/us/business/article/melinda-french-gates-bill-interview-ht27nw9p9.

Akyeampong, K. *et al.* (2023) Cost-effective approaches to improve global learning: What does recent evidence tell us are smart buys for improving learning in low- and middle-income countries?, London, Washington, DC, New York: FCDO, the World Bank, UNICEF, and USAID, https://documents.worldbank.org/en/publication/documents-reports/documentdetail/099420106132331608/idu0977f73d7022b1047770980c0c5a14598eef8.

Alami, I., Dixon, A.D., and Mawdsley, E. (2021) State capitalism and the new global D/development regime, *Antipode*, 53(5), 1294–1318, https://onlinelibrary.wiley.com/doi/full/10.1111/anti.12725.

Alarcon, S. and Verheije, T. (2024) *Smart Cities: A Blueprint for Urban Transformation in Latin America*, La Jolla, CA: Institute of the Americas.

Allen, K-A., Reardon, J., Crawford, J., and Walsh, L. (2022) The peer review system is broken. We asked academics how to fix it, *The Conversation*, https://theconversation.com/the-peer-review-system-is-broken-we-asked-academics-how-to-fix-it–187034.

Alvarez, J.M., Colmenarejo, A.B., Elobaid, A. *et al.* (2024) Policy advice and best practices on bias and fairness in AI, *Ethics and Information Technology*, 26 (31), https://link.springer.com/article/10.1007/s10676-024-09746-w.

Ames, M.G. (2019) *The Charisma Machine: The Life, Death, and Legacy of One Laptop per Child*, Cambridge, MA: MIT Press.

Amnesty International (2023) Saudi Arabia 2023, https://www.amnesty.org/en/location/middle-east-and-north-africa/middle-east/saudi-arabia/report-saudi-arabia/.

Andjelkovic, M. (2007) Internet Governance Forum: a commentary on the first meeting, International Institute for Sustainable Development, https://www.iisd.org/system/files/publications/infosoc_internet_governance_forum.pdf.

APC (2005) APC's recommendations to the WSIS on Internet governance, November 2025, https://dev-d10.apc.org/sites/default/files/apc_recommendations_ig_EN_0.pdf.

APC (2024a) Human rights, https://www.apc.org/en/our-work/themes/human-rights.

APC (2024b) Recapitulating of civil society feedback for the Global Digital Compact final version and call to action to incorporate observations, https://www.apc.org/en/pubs/recapitulating-civil-society-feedback-global-digital-compact-final-version-and-call-action.

Applequist, D. *et al.* (2024) An open letter to the United Nations, 1 July 2024, https://open-internet-governance.org/letter.

Araujo, W., Soares, D., and Carvalho, J. (2021) International rankings in e-Gov strategy formulation: Appraisal of their use and relevance, *IFIP EGOV-CeDEM-ePart Conference*, https://ceur-ws.org/Vol-3049/Paper17.pdf.

Arora, P. (2019) *The Next Billion Users: Digital Life Beyond the West*, Cambridge MA: Harvard University Press.

Arora, P. (2024) *From Pessimism to Promise: Lessons from the Global South on Designing Inclusive Tech*, Cambridge, MA: MIT Press.

Arruda, P.L. de, Andrade, M.L. de., Falcão, T., Barbosa, D.T., and Morgandi, M. (2021) *Auxílio Emergencial: Lessons from the Brazilian Experience Responding to COVID-19*, Washington, DC: World Bank, https://documents.worldbank.org/en/publication/documents-reports/documentdetail/099255012142121495/p1748361b302ee5718913146b11956610692e4faf5bc.

Arthur, C. (2012) *Digital Wars: Apple, Google, Microsoft & the Battle for the Internet*, London: Kogan Page.

ASEAN (2025) ASEAN Smart City Network, https://asean.org/our-communities/asean-smart-cities-network/.

Asgard, Z. (2025) No, your career isn't brilliant. Why Gen Z need to stop posturing, *The Times*, 31 March, https://www.thetimes.com/life-style/sex-relationships/article/gen-z-careers-posturing-boasting-achievements-gzn32htg2.

Autus (2025) About us, https://www.autus.org.uk/about_us.

Awde, N., Banes, D., and Banes, K. (2022) *Digital Assistive Technology: A Guide for People with Disabilities*, London: Millennium Community Solutions.

Baer, D. (2014) Mark Zuckerberg explains why Facebook doesn't 'move fast and break things' anymore, *Business Insider*, 2 May, https://www.businessinsider.com/mark-zuckerberg-on-facebooks-new-motto-2014–5?op=1.

Banerjee, B. (2021) Modern slavery is an enabling condition of global neoliberal capitalism: Commentary on modern slavery in business, *Business & Society*, 60(2), 415–419, https://journals.sagepub.com/doi/full/10.1177/0007650319898478.

Banks, K. (2013a) *The Rise of the Reluctant Innovator*, London: London Publishing Partnership.

Banks, K. (2013b) Reflections on a decade of Mobiles in Development, *Stability: International Journal of Security and Development*, 2(3), Art. 51, https://stabilityjournal.org/articles/10.5334/sta.cm.

Banks, K. (ed.) (2016) *Social Entrepreneurship and Innovation: International Case Studies and Practice*, London: Kogan Page.

Banks, K. (2022) *The Pursuit of Purpose: Part Memoir, Part Study – a Book About Finding Your Way in the World*, London: kiwanja.net.

Baraka, C. (2021) The failed promise of Kenya's smart city: African governments keep burning money planning smart cities by following a McKinsey blueprint. But will any actually be completed? *Rest of World*, https://restofworld.org/2021/the-failed-promise-of-kenyas-smart-city/.

Bartlett, J. (2015) *The Dark Web*, London: Windmill Books.

Baumann, B. (2021) Lessons learned from the NHS IT system failure, Panorama Consulting, https://www.panorama-consulting.com/nhs-it-system-failure/.

Bayliss-Smith, T. and Wanmali, S. (2009) *Understanding Green Revolutions: Agrarian Change and Development Planning in South Asia*, Cambridge: Cambridge University Press.

BBC (2022) *The Whistleblowers: Inside the UN*, BBC2, 21 June 2022 (Director: Ben Steele).

Beahm, G. (ed.) (2012) *The Boy Billionaire: Mark Zuckerberg: In His Own Words*, Evanston, IL: Agate.

Beardon, H., with Munyampeta, F., Rout, S., and Williams, G.M. (2004) ICT for development: Empowerment or exploitation? Learning from the *Reflect* ICTs project, London: Action Aid.

Beck, K. *et al.* (2001) The Agile Manifesto, Agile Alliance, http://agilemanifesto.org/.

Bedigan, M. (2024) A bromance for the ages: The timeline of Donald Trump and Elon Musk's tumultuous relationship, *The Independent*, 31 December https://www.independent.co.uk/news/world/americas/us-politics/trump-musk-friendship-history-b2672224.html.

Benson, V., McAlaney, J., Taylor, J., and Segumpan, R.G. (eds) (2024) *Handbook of Social Media Use: Online Relationships, Security, Privacy, and Society*, vol. 2, London: Academic Press.

Berners-Lee, T. (2025) *This Is for Everyone*, London: Pan Macmillan.

Bernick, M. (2024) The tech entrepreneurs of autism, *Forbes*, https://www.forbes.com/sites/michaelbernick/2024/06/04/the-tech-entrepreneurs-of-autism/.

Best, M.L. and Maclay, C.M. (2002) Community Internet access in rural areas: Solving the economic sustainability puzzle, in: Kirkman, G.S., Cornelius, P.K., Sachs, J.D., and Schwab, K. (eds), *The Global Information Technology Report 2001–2002: Readiness for the Networked World*, Oxford: Oxford University Press, 76–89.

Bhorat, H., Signé, L., Asmal, Z., Monnakgotla, J., and Rooney, C. (2023) *Digitalization and Digital Skills Gaps in Africa: An Empirical Profile*, Washington, DC: African Growth Initiative at Brookings, and Development Policy Research Unit.

BiologyInsights Team (2025) Neuroplasticity in children: Key factors and insights, https://biologyinsights.com/neuroplasticity-in-children-key-factors-and-insights/.

Bisits Bullen, P. (2022) What can international development learn from tech start-ups?, *tools4dev*, https://tools4dev.org/resources/what-can-international-development-learn-from-tech-start-ups/.

Blankenship, L. (1990) *Gurps Cyberpunk: High-Tech Low-Life Roleplaying*, Austin, TX: Steve Jackson Games.

Blavot, O., Romeri, M.J. and Ravenscroft, J. (eds) (2022) *History RePPPeated II: Why Public-Private Partnerships are not the solution*, Brussels: Eurodad, https://www.eurodad.org/historyrepppeated.

Bliss, T. and Smith, M. (2017) A brief history of Open Educational Resources, in: Jhangiani, R. and Biswas-Diener, R. (eds), *Open: The Philosophy and Practices that Are Revolutionizing Education and Science*, London: Ubiquity Press, 9–27.

Bloch, M. (1961) *Feudal Society*, Chicago: University of Chicago Press.

Bloch, M. (1967) *Land and Work in Medieval Europe*, Berkeley, CA: University of California Press.

Bloomberg (2025) Bloomberg Billionaires Index, 19 February 2025, https://www.bloomberg.com/billionaires/.

Blum, M. and Blum, L. (2021) A theoretical computer science perspective on consciousness, *Journal of Artificial Intelligence and Consciousness,* 8, 1–42, https://www.sciencedirect.com/science/article/pii/S2095809923001650.

BMJ (2024) Editorial: Selling NHS patient data, *BMJ*, 384, https://www.bmj.com/content/384/bmj.q420.

Bond (2020) Falling income, redundancies and programme cuts: Can NGOs survive the next two years?, https://www.bond.org.uk/news/2020/10/falling-income-redundancies-and-programme-cuts-can-ngos-survive-the-next-two-years.

Bound, K. and Ramalingam, B. (2016) Searching for the state of the art, in: Ramalingam, B. and Bound, K. (eds), *Innovation for International Development: Navigating the Paths and Pitfalls*, London: Nesta, 8–21, https://www.nesta.org.uk/report/innovation-for-international-development.

Bowen, H.R. (1953) *Social Responsibilities of the Businessman*, Iowa, IA: University of Iowa Press.

Bradley, K., Cartlidge, O., Perry, C., Eltis, D., Engerman, S.L., Richardson, D., and Drescher, S. (eds) (2011–2021) *The Cambridge History of Slavery*, Cambridge: Cambridge University Press.

Brandt, R.L. (2011) *One Click: Jeff Bezos and the Rise of Amazon.com*, London: Portfolio Penguin.

Brechenmacher, S. and Carothers, T. (eds) (2018) Examining civil society legitimacy, Carnegie Civic Research Network, https://carnegieendowment.org/posts/2018/05/examining-civil-society-legitimacy?lang=en.

Brown, R. (2010) *Gandhi's Spinning Wheel and the Making of India*, London: Routledge.

Bullinger, A.C., Haller, J., and Moeslein, K. (2009) Innovation mobs – unlocking the innovation potential of virtual communities, *AMCIS 2009 Proceedings*, 540, https://aisel.aisnet.org/cgi/viewcontent.cgi?article=1537&context=amcis2009.

Business Insider (2009) Mark Zuckerberg on innovation, https://www.businessinsider.com/mark-zuckerberg-innovation-2009–10.

Business Standard (2024) India's digital economy is growing 2.8x of GDP: Rajeev Chandrasekhar, *Business Standard*, 24 September, https://www.business-standard.com/industry/news/india-s-digital-economy-is-growing-2–8x-of-gdp-rajeev-chandrasekhar-124052001425_1.html.

Buskens, I. and Webb, A. (2014) *Women and ICT in Africa and the Middle East: Changing Selves, Changing Societies*, London: Zed.

CcHub (2025) Startups, https://cchub.africa/startups/.

CEB (2024) Total expenses, New York: UN Chief Executives Board for Co-ordination, https://unsceb.org/total-expenses.

Ceballos, F. Chittoor, J.P., and Ehrlich, P. (2006) *From the Ground Up: The Evolution of the Telecentre Movement*, Ottawa: IDRC, https://idl-bnc-idrc.dspacedirect.org/items/eac32204-a2fa-4a89–8036-b75ddf0ddf19.

Chamanara, S., Ghaffarizadeh, S.A., and Madani, K. (2023) The environmental footprint of Bitcoin mining across the globe: Call for urgent action, *Earth's Future*, 11, https://agupubs.onlinelibrary.wiley.com/doi/10.1029/2023EF003871.

Chambers, J. (2018) *Connecting the Dots: Lessons for Leadership in a Startup World*, New York: Hachette.

Chambers, S. and Kymlicka, W. (2002) *Alternative Conceptions of Civil Society*, Princeton, NJ: Princeton University Press.

Chance, L., Picketty, T., Saez, E., Zucman, G. *et al.* (2022) *World Inequality Report 2022*, World Inequality Lab, https://wir2022.wid.world/www-site/uploads/2022/01/Summary_WorldInequalityReport2022_English.pdf.

Chandhiramowuli, S., Taylor, A., Heitlinger, S., and Wang, D. (2023) Making data work count, *Proceedings of the ACM on Human-Computer Interaction*, 8 (CSCW1), No. 90, 1–26, https://dl.acm.org/doi/10.1145/3637367.

Chaney, P. and Jones, I.R. (eds) (2022) *Civil Society in an Age of Uncertainty*, Bristol: Bristol University Press.

Charles River Editors (2016) *Barons: The Lives and Careers of John D. Rockefeller, J.P. Morgan, Andrew Carnegie, and Cornelius Vanderbilt*, Scotts Valley, CA: CreateSpace Independent Publishing Platform.

Chen, Z. (2023) Ethics and discrimination in artificial intelligence-enabled recruitment practices, *Humanities and Social Science Communications*, 10, 567, https://www.nature.com/articles/s41599-023-02079-x.

Cheney-Lippold, J. (2017) *We Are the Data: Algorithms and the Making of Our Digital Selves*, New York: New York University Press.

China Development Brief (n.d.) Alibaba Foundation, https://chinadevelopmentbrief.org/ngos/alibaba-foundation/.

Chisnall, M. (2020) Digital slavery, time for abolition, *Policy Studies*, 41(5), 488–506, https://www.tandfonline.com/doi/full/10.1080/01442872.2020.1724926.

Chowdhury, F. (2018) Grade inflation: Causes, consequences and cure, *Journal of Education and Learning*, 7(6), 86–92, https://www.ccsenet.org/journal/index.php/jel/article/view/0/36791.

Christensen, C.M., Ojomo, E., and Dillon, K. (2019) *The Prosperity Paradox: How Innovation Can Lift Nations Out of Poverty*, New York: HarperBusiness.

Chu, T. (2020) *The History of Chinese Feudal Society*, London: Routledge.

Cisco (2004) Company profile, https://www.cisco.com/c/dam/en_us/about/ac49/ac20/downloads/annualreport/ar2004/pdf/ar_2004_profile_and_offices.pdf.

Cisco (2024) FY24 Purpose Report: Celebrating 40 years of impact, https://www.cisco.com/site/us/en/about/purpose/fy24-report/index.html.

Clarke, T. (2017) *International Corporate Governance: A Comparative Approach*, 2nd edition, London: Routledge.

Cloninger, C.R. (1994) *The Temperament and Character Inventory (TCI): A Guide to Its Development and Use*, St. Louis, MA: Washington University.

Clough, E. (2015) Effective altruism's political blindspot, *Boston Review*, https://www.bostonreview.net/articles/emily-clough-effective-altruism-ngos/.

Clynes, M.E. and Kline, N.S. (1960) Cyborgs and space, *Astronautics*, September, 26–27, 74–76, https://fermatslibrary.com/s/cyborgs-and-space.

Code for America (2018) Talent Initiative, Impact Report, https://codeforamerica.org/impact/2018/index.html.

Coeckelbergh, M. (2024) *Why AI Undermines Democracy and What to Do About It*, Cambridge: Polity Press.

COL (2022) *Open Educational Resources in the Commonwealth 2021*, Vancouver: Commonwealth of Learning, https://oasis.col.org/server/api/core/bitstreams/61d83c16–3f77–414d-99b0–5d4f9471c7d4/content.

Collier, P. (2007) *The Bottom Billion*, Oxford: Oxford University Press.

Colman, R. (2021) *What Really Counts: The Case for a Sustainable and Equitable Economy*, New York: Columbia University Press.

Companiesmarketcap (2024) Largest tech companies by market cap, https://companiesmarketcap.com/gbp/tech/largest-tech-companies-by-market-cap/.

Comte, A. (1875) *System of Positive Polity*, London: Longmans, Green & Co (first published in French, 1851–1854).

Cooper, R. (2018) What is Civil Society, its role and value in 2018?, *K4D: Knowledge, Evidence and Learning for Development*, London: DFID, https://assets.publishing.service.gov.uk/media/5c6c2e74e5274a72bc45240e/488_What_is_Civil_Society.pdf.

Cornish, P. (ed.) (2021) *The Oxford Handbook of Cyber Security*, Oxford: Oxford University Press.

Cousens, E. (2024) The Summit of the Future – my thoughts on what's next, United Nations Foundation, https://unfoundation.org/blog/post/the-summit-of-the-future-my-thoughts-on-whats-next/.

Cowley, C. (2014) *Moral Responsibility*, London: Routledge.

Crabbe, J., Barker, G.R., Wong, L., and Zhou, J. (2024) The fragility of the ocean: From coral reef protection to deep-sea mining, in: Monaco, E. and Abe, M. (eds), *Sustainable Development Across Pacific Islands*, Singapore: Springer, 187–202.

Cragg, W., Schwartz, M.S., and Weitzner, D. (eds) (2009) *Corporate Social Responsibility*, Farnham: Ashgate.

Crary, A. (2021) Against 'Effective Altruism', *Radical Philosophy*, 210, 33–43, https://www.radicalphilosophy.com/wp-content/uploads/2021/07/rp210_crary.pdf.

Crawford, K. (2024) Generative AI's environmental costs are soaring – and mostly secret, *Nature*, 626, 693, https://www.nature.com/articles/d41586-024-00478-x.

Creative Commons (n.d.) A history of Creative Commons, https://creativecommons.org/timeline/.

CTO (2014) Statement on the role of ICTs in the post-2015 Development Goals, London: CTO, https://unwins.info/wp-content/uploads/2023/04/Post-2015-goals-final.pdf.

Cunha, P.R. da, Soja, P., and Themistocleous, M. (2021) Blockchain for development: A guiding framework, *Information Technology for Development*, 27(3), 417–438, https://www.tandfonline.com/doi/pdf/10.1080/02681102.2021.1935453.

Curtis, M. (2016) *Gated Development: Is the Gates Foundation always a force for good?*, 2nd edn. London: Global Justice Now, https://www.globaljustice.org.uk/wp-content/uploads/2016/06/gjn_gates_report_june_2016_web_final_version_2.pdf.

D'Alisa, G., Demaria, F., and Kallis, G. (eds) (2015) *Degrowth: A Vocabulary for a New Era*, London: Routledge.

Deardon, A., Rizvi, H., and Gupta, S. (2010) Roles and responsibilities in agile ICT for development, *IHCI'10: Proceedings of the 2010 International Conference on Interaction Design & International Development*, 19–30, https://dl.acm.org/doi/10.5555/2227347.2227350.

Deckelmann, S. (2023) Wikipedia's value in the age of generative AI, *Wikimedia Foundation*, https://wikimediafoundation.org/news/2023/07/12/wikipedias-value-in-the-age-of-generative-ai/.

Denardis, L., Cogburn, D., Levinson, N.S., and Musiani, F. (eds) (2020) *Researching Internet Governance: Methods, Frameworks, Futures*, Cambridge, MA: MIT Press.

Dennis, M., Spiegler, B.J., Juranek, J.J., Bigler, E.D., Snead, O.C., and Fletcher, J.M. (2013) Age, plasticity, and homeostasis in childhood brain disorders, *Neuroscience and Biobehavioral Reviews*, 37(10 Pt 2), 2760–2773, https://pubmed.ncbi.nlm.nih.gov/24096190/.

Dercon, S. (2023) *Gambling on Development: Why Some Countries Win and Others Lose*, London: Hurst.

Dercon, S. (2025) Addressing the problem of best buys in politics: The political economy of education research, https://www.wwhge.org/resources/addressing-the-problem-of-best-buys-in-politics-the-political-economy-of-education-research/.

DESC (2021) Digital-Environment System Coalition: DESC conceptual framework, ICT4D Collective, https://ict4d.org.uk/wp-content/uploads/2021/09/conceptual-framework-v3.pdf.

De Souza, S.P. (2025) Can data justice be global? Exploring the practice of digital rights, and the search for cognitive data justice, *Information, Communication & Society*, 1006–1022, https://www.tandfonline.com/doi/pdf/10.1080/1369118X.2025.2471889.

Devex (2023) The top 10 foundations funding development, Devex, https://pages.devex.com/rs/685-KBL-765/images/the-top-10-foundations-funding-development.pdf.

Devlin, K. (2018) *Turned On: Science, Sex and Robots*, London: Bloomsbury.

DFID (2004) *The Rough Guide to a Better World and How You Can Make a Difference*, London: Rough Guides.

DFID (2019) Programme completion report summary of infoDev 112024, London: DFID.

DIAL (2025) Case study: The Principles for Digital Development: a decade of transforming digital development practice (2014–2024), Washington, DC: Digital Impact Alliance.

Digital News Asia (2013) Malaysian seaweed farmer wins ITU-MCMC 'Digital Icon Award', *Digital News Asia*, 22 November, https://www.digitalnewsasia.com/digital-economy/malaysian-seaweed-farmer-wins-itu-mcmc-digital-icon-award.

digitalprinciples.org (2024) The Principles for Digital Development: A compass for those working to promote sustainable and inclusive development in today's complex digital landscape, https://digitalprinciples.org/.

digwatch (2024) Internet Governance Forum in 2024: Process, issues, and actors, https://dig.watch/processes/igf-process.

Dijkers, J.J.G., Overbeek, S.J., and España, S. (2018) Improving ICT4D projects with Agile software development, *Proceedings of the 5th International Symposium "Perspectives on ICT4D"*, (P-ICT4D 2018), 2120, Amsterdam: CEUR-WS, https://ceur-ws.org/Vol-2120/paper4.pdf.

Dimick, M., Rueda, D., and Stegmueller, D. (2016) The altruistic rich? Inequality and other-regarding preferences for redistribution, *Quarterly Journal of Political Science*, 11(4), 385–439, https://rueda.web.ox.ac.uk/files/dimickruedastegmueller2016pdf.

Dinesson, K.E. (2024) Mr Bates vs The Post Office: ITV, January 2024, four-part TV drama written by Gwyneth Hughes, directed by James Strong, with ensemble cast led by Toby Jones and Monica Dolan, *Law and Humanities*, 18(1), 137–142, https://www.tandfonline.com/doi/full/10.1080/17521483.2024.2323357.

DIRCO (2025) Alleged South African victims of human trafficking activity successfully repatriated from Myanmar, Department of International Relations and Cooperation, Republic of South Africa, https://dirco.gov.za/alleged-south-african-victims-of-human-trafficking-activity-successfully-repatriated-from-myanmar/.

Doerflinger, J. and Deardon, A. (2013) Evolving a software development methodology for commercial ICTD projects, *Information Technology and International Development*, 9(3), 43–60, https://itidjournal.org/index.php/itid/article/view/1089.html.

DOJ (2023) Justice Department sues Google for monopolizing digital advertising technologies, https://www.justice.gov/opa/pr/justice-department-sues-google-monopolizing-digital-advertising-technologies.

DOJ (2024) Justice Department sues Apple for monopolizing smartphone markets, https://www.justice.gov/opa/pr/justice-department-sues-apple-monopolizing-smartphone-markets.

Dolinger, J. (2016) The failure of the Universal Declaration of Human Rights, *University of Miami Inter-American Law Review*, 47(2/4), 164–199, https://repository.law.miami.edu/umialr/vol47/iss2/4/.

Donne, J. (1624) *Devotions Upon Emergent Occasions, and Several Steps in My Sickness*, London.

Dostoyevsky, F. (2021) *Notes from the Underground*, Salt Lake City, UT: Project Gutenberg (first published in 1864). https://www.gutenberg.org/cache/epub/600/pg600-images.html.

DOT (2022) Creating transformative youth digital spaces, Dar es Salaam: DOT Tanzania, https://dotrust.org/wp-content/uploads/2024/05/Transformative-Youth-Digital-Spaces-Report-Insights-from-the-Tanzania-Youth-Digital-Summit−2022.pdf.

DOT (2023) Enabling ecosystems for young women's digital inclusion, Kigali: DOT Rwanda, https://dotrust.org/wp-content/uploads/2024/05/Enabling-Ecosystems-for-Young-Womens-Digital-Inclusion.pdf.

DSIT and DBEIS (2023) UK Innovation Strategy: Leading the future by creating it, https://www.gov.uk/government/publications/uk-innovation-strategy-leading-the-future-by-creating-it/uk-innovation-strategy-leading-the-future-by-creating-it-accessible-webpage#introduction-why-do-we-need-an-innovation-strategy.

Duff, R.A. (1998) Responsibility, in: *Routledge Encyclopedia of Philosophy*, London: Routledge.

Dunford, M. (2021) China's Belt and Road Initiative and its implications for global development, *Acta Via Serica*, 6(1), 91–118, https://koreascience.or.kr/article/JAKO202118053301312.page.

Durand, C. (2024) *How Silicon Valley Unleashed Techno-Feudalism: The Making of the Digital Economy*, trans. D. Broder, London: Verso.

Dutton, W.H. (2013) The social shaping of digital research, *International Journal of Social Research Methodology*, 16(3), 177–195, https://www.tandfonline.com/doi/full/10.1080/13645579.2013.774171.

Duval, M. (2025) Cape Town mom in court for allegedly prostituting her teen daughter for milk money, https://iol.co.za/capeargus/news/2025-04-29-cape-town-mother-in-court-for-allegedly-selling-daughter-for-milk-money/.

Easterly, W. (2006) *The White Man's Burden: Why the West's Efforts to Aid the Rest Have Done So Much Ill and So Little Good*, Oxford: Oxford University Press.

Economist Intelligence Unit (2025) *Democracy Index 2024: What's Wrong with Representative Democracy?*, London: Economist Intelligence Unit, https://www.eiu.com/n/global-themes/democracy-index/.

Edwards, M. (2009) Gates, Google, and the ending of global poverty: Philanthrocapitalism and international development, *The Brown Journal of World Affairs*, 15(2), 35–42, https://www.jstor.org/stable/24590840.

Effective Altruism (n.d.) What is effective altruism?, https://www.effectivealtruism.org/articles/introduction-to-effective-altruism.

Egbetokun, A. (2023) Business incubators in Africa: A review of the literature, *Innovation and Development*, 15(2), 233–260, https://www.tandfonline.com/doi/pdf/10.1080/2157930X.2023.2295090.

Eid, N. (2017) *Disability and bridging the digital divide*, Ruh Global Impact, https://www.ruhglobal.com/wp-content/uploads/2017/01/Nabil-Eid-from-Syria-Middle-East-ICT-accessibility-for-persons-with-disabilities.pdf.

Elder, L., Emdon, H., Fuchs, R., and Petrazzini, B. (2013) *Connecting ICTs to Development: The IDRC Experience*, Ottawa: Anthem.

Emanuel, P., Walper, S., DiEuliis, D., Klein, N., Petro, J.B., and Giordano, J. (2019) Cyborg Soldier 2050: Human/machine fusion and the implications for the future of DOD, US Army DEVCOM, CCDC CBC-TR-1599, https://apps.dtic.mil/sti/html/trecms/AD1083010/index.html.

Erasmus, J. (2008) Science on the up in Mozambique, *Brand South Africa*, https://brandsouthafrica.com/111866/innovations/mozambique-science-parks-101108/.

Ericsson (n.d.) The Millennium Villages Project, https://www.ericsson.com/en/about-us/sustainability-and-corporate-responsibility/digital-inclusion/access-to-education/the-millennium-villages-project.

Eskreis-Winkler, L. and Fishbach, A (2019) Not learning from failure – the greatest failure of all, *Psychological Science*, 30(12), 1733–1744, https://journals.sagepub.com/doi/abs/10.1177/0956797619881133.

European Commission (2022) The Digital Services Act, https://commission.europa.eu/strategy-and-policy/priorities-2019–2024/europe-fit-digital-age/digital-services-act_en.

European Network for Rural Development (2021) Smart villages, https://ec.europa.eu/enrd/smart-and-competitive-rural-areas/smart-villages_en.html.

EvalCommunity (n.d.) Finding the right budget allocation for Monitoring and Evaluation (M&E), https://www.evalcommunity.com/donors-and-nonprofit/budget-allocation-for-monitoring-and-evaluation-me/.

Farrell, G., Isaacs, S., and Trucano, M. (2007) *The NEPAD e-Schools Demonstration Project: A Work in Progress. A Public Report*, Vancouver: COL, and Washington DC: *info*Dev, World Bank, https://documents1.worldbank.org/curated/en/208421468140054464/pdf/408800NEPAD0e11blic0Report01PUBLIC1.pdf.

FCDO (n.d.) Global Innovation Fund, https://devtracker.fcdo.gov.uk/programme/GB-1–202604/summary.
Feenberg, A. (1991) *Critical Theory of Technology*, Cambridge: Cambridge University Press.
Fernando, J. (2024) What is CSR? Social responsibility explained, *Investopedia*, https://www.investopedia.com/terms/c/corp-social-responsibility.asp.
Ferrara, E. (2024) Fairness and bias in Artificial Intelligence: A brief survey of sources, impacts, and mitigation strategies, *Sci*, 6(1), 3, https://www.mdpi.com/2413-4155/6/1/3.
Finlay, A. and Betancourt, V. (2024) How the Global Digital Compact is implemented will determine our future, Association for Progressive Communications, https://www.apc.org/en/news/how-global-digital-compact-implemented-will-determine-our-digital-future.
Fleck, A. (2023) The cost of mobile Internet around the world, *Statista*, https://www.statista.com/chart/17247/the-average-cost-of-mobile-data-in-selected-countries/.
Fletcher-Watson, S. and Happé, F. (2019) *Autism: A New Introduction to Psychological Theory and Current Debate*, New York: Routledge.
Flickenger, R. (ed.) (2006) *Wireless Networking in the Developing World: A Practical Guide to Planning and Building Low-Cost Telecommunications Infrastructure*, Limehouse Book Sprint Team.
Floridi, L. (2023) *The Ethics of Artificial Intelligence: Principles, Challenges, and Opportunities*, Oxford: Oxford University Press.
Flyvbjerg, B., Budzier, A., Lee, J.S., Keil, M., Lunn, D., and Bester, D.W. (2022) The empirical reality of ITB project cost overruns: Discovering a power-law distribution, *Journal of Management Information Systems*, 39(3), 607–639, https://www.jmis-web.org/articles/1579.
Forbes (2024a) The world's real-time billionaires, https://www.forbes.com/real-time-billionaires.
Forbes (2024b) The richest crypto and bitcoin billionaires in the world 2024, https://www.forbes.com/sites/johnhyatt/2024/04/02/the-richest-crypto-and-bitcoin-billionaires-in-the-world-2024/.
Forbes (2025) Bill Gates: Cochair, Bill & Melinda Gates Foundation, https://www.forbes.com/profile/bill-gates/.
Fortune Business Insights (2024) Technology: Data analytics market, https://www.fortunebusinessinsights.com/data-analytics-market–108882.
Frances, A. (2013) *Saving Normal: An Insider's Revolt Against Out-of-Control Psychiatric Diagnosis, DSM-5, Big Pharma, and the Medicalization of Ordinary Life*, New York: HarperCollins.
Frances, A. (2025) Autism rates have increases 60-fold. I played a role in that, *New York Times*, 23 June, https://www.nytimes.com/2025/06/23/opinion/why-autism-rates-increased.html.
Francis, S. (2024) NHS data sale can fuel tech boom – Blair and Hague, *BBC News* 26 January, https://www.bbc.co.uk/news/uk-politics–68094059.
Free Software Foundation (n.d.) What is GNU, https://www.gnu.org/.
Freeman, H. (2025) We've fetishised mental illness, so we all want it, *The Sunday Times*, 6 April, https://www.thetimes.com/article/e06e41bd-0371–4bd5-b2ae-589e52b23682.
Fridman, L. (2023) Transcript for Jeff Bezos: Amazon and Blue Origin. Lex Fridman Podcast #405, https://lexfridman.com/jeff-bezos-transcript/.
Fu, X., McKern, B., and Chen, J. (eds) (2021) *The Oxford Handbook of China Innovation*, Oxford: Oxford University Press.
Fuchs, C. (2023) *Digital Ethics: Media, Communication and Society*, vol. Five, London: Routledge.
Fuchs, C. (2024) *Media, Economy and Society: A Critical Introduction*, Abingdon: Routledge.

Gaballa, M. (2024) Looking beyond the Global Digital Compact, Internet Society, https://www.internetsociety.org/blog/2024/09/looking-beyond-the-global-digital-compact/.

Gallagher, W. (2011) *New: Understanding or Need for Novelty and Change*, New York: Penguin.

Ganamotse, G.N., Samuelsson, M., and Mphela, T. (2017) The emerging properties of business accelerators: The case of Botswana, Namibia and Uganda Global Business Labs, *Journal of Entrepreneurship and Innovation in Emerging Economies*, 3(1), 16–40, https://journals.sagepub.com/doi/10.1177/2393957516684469.

Gates, B. (2025) *Source Code_My Beginnings*, London: Allen Lane.

GDS and GDDO (2012) Government Design Principles, https://www.gov.uk/guidance/government-design-principles.

GeeksforGeeks (2024) Guidelines in Human Computer Interface (HCI), https://www.geeksforgeeks.org/guidelines-in-human-computer-interfacehci/#shneidermans-eight-golden-rules.

Gentzoglanis, A. and Henten, A. (eds) (2010) *Regulation and the Evolution of the Global Telecommunications Industry*, Cheltenham: Edward Elgar.

Giga (n.d.) Finance, https://giga.global/finance/.

GIZ (2012) *IT Sector Promotion in Developing and Emerging Countries*, Bonn: GIZ.

Google (n.d.) From the garage to the Googleplex, https://about.google/company-info/our-story/.

Gordon, P., Harries, J., Fischer, M., and Plamping, D. (1998) *Projectitis: Spending Lots of Money and the Trouble of Project Bidding*, London: The Kings Fund, https://archive.kingsfund.org.uk/concern/published_works/000019942?locale=fr

Graham, M. (ed.) (2019) *Digital Economies at Global Margins*, Cambridge, MA: MIT Press.

Gramsci, A. (1971) *Prison Notebooks*, New York: International Publishers (written 1929–1935).

Green, D. (2013) The idealist: A brilliant, gripping, disturbing portrait of Jeffrey Sachs, in: *From Poverty to Power*, Oxfam, https://frompoverty.oxfam.org.uk/the-idealist-a-brilliant-gripping-disturbing-portrait-of-jeffrey-sachs/.

Gross, A.G. (2010) Systematically distorted communication: An impediment to social and political change, *Informal Logic*, 30(4), 335–360, https://informallogic.ca/index.php/informal_logic/article/view/1188.

GSMA (2024a) *The State of Mobile Internet Connectivity Report 2024*, London: GSMA, https://www.gsma.com/r/wp-content/uploads/2024/10/The-State-of-Mobile-Internet-Connectivity-Report−2024.pdf.

GSMA (2024b) GSMA Ministerial Programme: The centre of policy debate, London: GSMA, https://www.mwcbarcelona.com/ministerial.

GSMA (2024c) *The Mobile Gender Gap Report 2024*, London: GSMA, https://www.gsma.com/r/wp-content/uploads/2024/05/The-Mobile-Gender-Gap-Report−2024.pdf.

GSMA (2024d) *The Mobile Economy Sub-Saharan Africa 2024*, London: GSMA, https://www.gsma.com/solutions-and-impact/connectivity-for-good/mobile-economy/sub-saharan-africa.

GSMA (2025) *State of the Industry Report: Mobile Money 2025*, London: GSMA, https://www.gsma.com/sotir/.

Gupta, P. (2024) Why both businesses and scammers love India's payment system, *BBC News*, https://www.bbc.co.uk/news/articles/c288m1km01po.

Guterres, A. (2020a) Nelson Mandela Annual Lecture 2020. Tackling inequality: a new social contract for a new era, https://www.un.org/en/coronavirus/tackling-inequality-new-social-contract-new-era.

Guterres, A. (2020b) Quotation introducing the Secretary General's Roadmap for Digital Cooperation, https://www.un.org/en/content/digital-cooperation-roadmap/.

Gyan, N.B. (2016) *The Web, Speech Technologies and Rural Development in West Africa: An ICT Approach*, Amsterdam: Vrije Universiteit Amsterdam.

Habermas, J. (1974) *Theory and Practice*, London: Heinemann.

Habermas, J. (1978) *Knowledge and Human Interests*, 2nd edn, London: Heinemann.
Hägerstrand, T. (1953) *Innovationsförloppet ur korologisk synpunkt*, Lund: Lund University.
Haidt, J. and Allen, N. (2020) Scrutinizing the effects of digital technology on mental health, *Nature*, 578, 226–227, https://www.nature.com/articles/d41586-020-00296-x.
Haikin, M. (2013) Reflections on applying Agile in developing countries, https://matthaikin.com/wp-content/uploads/2013/03/agile-blarticle-part−11.pdf.
Haji, I. (2023) *Obligation and Responsibility*, Oxford: Oxford University Press.
Hamilton, L. (2003) 'Civil Society': critique and alternative, in: Laxer, G. and Halperin, S. (eds), *Global Civil Society and Its Limits*, Basingstoke: Palgrave Macmillan, 63–81.
Hamilton, S.N. (1997) The Cyborg, 11 years later: The not-so-surprising half-life of the *Cyborg-Manifesto*, *Convergence: The International Journal of Research into New Media Technologies*, 3(2), 104–120, https://journals.sagepub.com/doi/abs/10.1177/135485659700300211?download=true.
Haraway, D. (1991) *Simians, Cyborgs, and Women: The Reinvention of Nature*, New York: Free Association Books.
Harris, R.W. (2016) How ICT4D research fails the poor, *Information Technology for Development*, 22(1), 177–192, https://www.tandfonline.com/doi/full/10.1080/02681102.2015.1018115.
Hasell, J. (2023) Measuring inequality: What is the Gini coefficient?, https://ourworldindata.org/what-is-the-gini-coefficient.
Hatcher, J. (1981) English serfdom and villeinage: Towards a reassessment, *Past & Present*, 90, 33–39, https://www.jstor.org/stable/650715.
Heeks, R. (ed.) (1995) *Technology and Developing Countries: Practical Applications, Theoretical Issues*, London: Routledge.
Heikkilä, M. (2021) António Guterres' role questioned in UN harassment case, *Politico*, 19 May, https://www.politico.eu/article/antonio-guterres-harassment-united-nations-fabrizio-hochschild-drummond/.
Hempel, J. (2018) What happened to Facebook's grand plan to wire the world?, *Wired*, https://www.wired.com/story/what-happened-to-facebooks-grand-plan-to-wire-the-world/.
Herzog, S.M. and Hertwig, R. (2025) Boosting: Empowering citizens with behavioral science, *Annual Review of Psychology*, 76, 851–881, https://www.annualreviews.org/content/journals/10.1146/annurev-psych-020924-124753.
Hickel, J. (2020) *Less Is More: How Degrowth Will Save the World*, London: Penguin Random House.
Hilton, R.H. (1965) Freedom and villeinage in England, *Past & Present*, 31, 3–19, https://academic.oup.com/past/article-abstract/31/1/3/1468667?login=false.
Hiltzik, M. (2020) *Iron Empires: Robber Barons, the Railroads, and the Making of Modern America*, Stroud: Amberley Publishing.
Hochschild, F. (2021) Tweet, X, https://x.com/HochschildF/status/1352789899938824192.
Holden, K. and Harsh, M. (2024) On pipelines, readiness and annotative labour: Political geographies of AI and data infrastructures in Africa, *Political Geography*, 113, 103150, https://www.sciencedirect.com/science/article/pii/S0962629824000994.
Hollow, D. (2015) Monitoring and evaluation of ICT4D, in: Ang, P.H. and Mansell, R. (eds), *The International Encyclopedia of Digital Communication and Society*, Chichester: Wiley.
Hollow, D. and Jefferies, K. (2022) How EdTech can be used to help address the Global Learning Crisis: A challenge to the sector for an evidence-driven future, EdTech Hub, https://doi.org/10.53832/edtechhub.0115.
Horkheimer, M. (1972) *Critical Theory*, New York: Seabury Press.
Howell, C. and West, D.M. (2016) The Internet as a human right, *Brookings Commentary*, 7 November, https://www.brookings.edu/articles/the-internet-as-a-human-right/.
Hoxtell, W. and Nonhoff, D. (2019) *Internet Governance: Past, Present and Future*, Berlin: Global Public Policy Institute, Konrad Adenauer Stiftung.

Hund, E. (2023) *The Influencer Industry: The Quest for Authenticity on Social Media*, Princeton, NJ: Princeton University Press.
Hunsings, J. and Senft, T. (eds) (2023) *The Social Media Handbook*, London: Routledge.
IAMAI (Internet and Mobile Association of India) and Kantar (2023) Internet in India 2023, https://datareportal.com/reports/digital-2025-india.
IEA (2022) *Africa Energy Outlook: Key Findings*, Paris: IEA, https://www.iea.org/reports/africa-energy-outlook-2022/key-findings.
IEA (2024) *SDG7: Data and Projections*, Paris: IEA, https://www.iea.org/reports/sdg7-data-and-projections.
IEEE (2024) IEEE connecting the unconnected: Solutions to the digital divide: moving toward a more equitable future, https://ctu.ieee.org/solutions-to-the-digital-divide-moving-toward-a-more-equitable-future/.
IEG (2011) *Capturing Technology for Development: An Evaluation of World Bank Group Activities in Information and Communication Technologies*, Washington, DC: The Independent Evaluation Group, The World Bank Group.
IGF (2024) IGF 2024 Annual Meeting Summary Report: Nineteenth Meting of Internet Governance Forum, 15–19 December 2024, Riyadh, Saudi Arabia, https://intgovforum.org/en/filedepot_download/305/28529.
Ignite (2024) Who we are, https://ignite.org.pk/who-we-are/.
iHub (2024) About us, https://ihub.co.ke/about-us/.
ILO, Walk Free and IOM (2022) *Global Estimates of Modern Slavery: Forced Labour and Forced Marriage*, Geneva: ILO, Walk Free and IOM, https://www.walkfree.org/reports/global-estimates-of-modern-slavery-2022/.
IMD (2025) World digital competitiveness ranking, Lausanne: IMD, https://www.imd.org/centers/wcc/world-competitiveness-center/rankings/world-digital-competitiveness-ranking/.
InCountry (2024) China's digital data sovereignty laws and regulations, *InCountry*, https://incountry.com/blog/chinas-digital-data-sovereignty-laws-and-regulations/.
*info*Dev (2005) *The* info *Dev Global Network of Business Incubators*, Washington, DC: IBRD/World Bank, https://documents.worldbank.org/en/publication/documents-reports/documentdetail/625921468158729636/the-infodev-global-network-of-business-incubators.
Inglis, T. (1997) Empowerment and emancipation, *Adult Education Quarterly*, 48(1), 3–17, https://journals.sagepub.com/doi/abs/10.1177/074171369704800102.
Intel (n.d.) *Reassessing ICTs and Development: The Social Forces of Consumption*, Santa Clara, CA: Intel, http://scott.mainzone.com/pubs/10-reassessing-icts-and-development.pdf.
International Energy Agency (2025) *Energy and AI*, Paris: International Energy Agency, https://www.iea.org/reports/energy-and-ai/.
Internet Watch Foundation (2023) *IWF Annual Report 2023*, Cambridge: Internet Watch Foundation, https://www.iwf.org.uk/annual-report-2023/.
Isaacson, W. (2015) *Steve Jobs: The Exclusive Biography*, London: Abacus.
Isaacson, W. (2025) *Elon Musk*, New York: Simon & Schuster.
ISTI (2024) India Science, Technology and Innovation Portal, https://www.indiascienceandtechnology.gov.in/.
ITCILO (2021) *From 4our To 4orty 2wo: Leaving No One Behind*, Turin: ITCILO.
ITU (2000) ITU Telecommunication Indicators Update – Uganda, April–June 2000, https://www.itu.int/ITU-D/ict/cs/uganda/material/CountryProfileUGA.pdf.
ITU (2005) *Digital Reach*, Geneva: ITU.
ITU (2006) World Summit on the Information Society: Geneva 2003 – Tunis 2005, https://www.itu.int/net/wsis/basic/index.html, updated 16 November 2006.
ITU (2011) *ICT and Telecommunications in Least Developed Countries: Review of Progress Made During the Decade 2000–2010*, Geneva: ITU, https://www.itu.int/pub/D-LDC-ICTLDC.2000.2010–2011.

ITU (2015) WSIS-SDG Matrix: Linking WSIS Action Lines with Sustainable Development Goals, Geneva: ITU, https://www.itu.int/net4/wsis/sdg/.

ITU (2021) Digital technologies to achieve the UN SDGs, https://www.itu.int/en/mediacentre/backgrounders/Pages/icts-to-achieve-the-united-nations-sustainable-development-goals.aspx.

ITU (2022) Achieving universal and meaningful digital connectivity in the decade of action: Aspirational targets for 2030, ITU, https://www.itu.int/itu-d/meetings/statistics/wp-content/uploads/sites/8/2022/04/UniversalMeaningfulDigitalConnectivityTargets2030.pdf.

ITU (2023) Partner2Connect Digital Coalition Annual Report 2023, Geneva: ITU, https://s46824.pcdn.co/partner2connect/wp-content/uploads/2025/01/P2C-Annual-Report–2023.pdf.

ITU (2024) Measuring digital development: Facts and figures 2024, Geneva: ITU, https://www.itu.int/en/ITU-D/Statistics/Pages/facts/default.aspx?os=a&ref=app.

ITU (2025) AI for Good, https://aiforgood.itu.int/.

ITU (n.d.) Universal and meaningful connectivity, https://www.itu.int/itu-d/sites/projectumc/home/aboutumc/.

ITU and UNDP (2023) *SDG Digital Acceleration Agenda*, Geneva and New York: ITU and UNDP, https://www.undp.org/publications/sdg-digital-acceleration-agenda.

Jack Ma Foundation (n.d.) About, https://www.jackmafoundation.org.cn/about/.

Jhangiani, R. and Biswas-Diener, R. (eds) (2017) *Open: The Philosophy and Practices that Are Revolutionizing Education and Science*, London: Ubiquity.

Joseph, S. (2022) The rundown: Google, Meta and Amazon are on track to absorb more than 50% of all ad money in 2022, *Digiday*, https://digiday.com/marketing/the-rundown-google-meta-and-amazon-are-on-track-to-absorb-more-than-50-of-all-ad-money-in-2022/.

Kaffenberger, M., Melville, S., and Agarwal, M. (2025) The benefits of foundational learning to individuals and society: A review of the evidence, *What Works Hub for Global Education, Synthesis Brief*, https://doi.org/10.35489/BSG-WhatWorksHubforGlobalEducation-RI_2025/003.

Kaspersky (2020) Empowering innovations: Create a work environment to drive positive change, https://media.kasperskydaily.com/wp-content/uploads/sites/92/2020/10/14181928/Kaspersky-Empowering-Innovations-Final.pdf.

Kaushik, D. (2021) Cyberspace as a global commons, *South Asia Journal*, https://southasia-journal.net/cyberspace-as-a-global-commons/.

Kelleher, K. (2013) Here comes Mark Zuckerberg's knowledge economy, *Fortune*, 5 November, https://fortune.com/2013/11/05/here-comes-mark-zuckerbergs-knowledge-economy/.

Kenny, S., Taylor, M., Onyx, J., and Mao, M. (2016) *Challenging the Third Sector: Global Prospects for Active Citizenship*, Bristol: Policy Press.

Keshvani-Ham, A. and Tiwari, A. (2025) India's cybercrime epidemic: Victims lose life savings to fake police, *The Times*, 2 August, https://www.thetimes.com/article/75214931-4257-4dcf-8a7b-436e9d9b8c37.

King, H. (2022) Elon Musk opens up on how Asperger's has impacted his life, *Axios*, https://www.axios.com/2022/04/15/elon-musk-aspergers-syndrome.

Kitchin, R. (2014) *The Data Revolution: Big Data, Open Data, Data Infrastructures and Their Consequences*, London: Sage.

Kitchin, R. (2023) *Digital Timescapes: Technology, Temporality and Society*, Cambridge: Polity Press.

Klein, M. (1997) *The Life and Legend of Jay Gould*, Baltimore, MD: Johns Hopkins University Press.

Klick, J. (2021) Big Tech's Robber Barons, *Regulation*, Fall, 2629, https://www.cato.org/regulation/fall-2021/big-techs-digital-robber-barons#.

KNBS and FSD (2021) 2021 finaccess household survey, Nairobi: Kenya National Bureaus of Statistics, https://www.fsdkenya.org/wp-content/uploads/2021/12/FinAccess-2021-Infographic.pdf.

Koch, D-J. (2024) *Foreign Aid and its Unintended Consequences*, London: Routledge.

Konza (2024) Konza Technopolis history, https://konza.go.ke/history/.

Korsheed, M.S., Al-Fawzan, M.A., and Al-Hargan, A. (2014) Promoting techno-entrepreneurship through incubation: An overview at BADIR program for technology incubators, *Innovation*, 16(2), 238–249, https://www.tandfonline.com/doi/abs/10.1080/14479338.2014.11081985.

Kosmyna, N., Hauptmann, E., Yuan, Y.T., Situ, J., Liao, X-H., Beresnitzkl, A.V., Braunstein, I., and Maes, P. (2025) Your brain on ChatGPT: Accumulation of cognitive debt when using an AI assistant for essay writing tasks, https://arxiv.org/pdf/2506.08872v1.

Kotecha, S. (2022) UN insists it is tackling sexism 'head on' amid new allegations, BBC News, 15 July, https://www.bbc.co.uk/news/world-us-canada−62159610.

KPMG (2014) Monitoring and Evaluation in development report: A KPMG International Development Assistance (IDAS) practice survey, London: KMPG, https://assets.kpmg.com/content/dam/kpmg/pdf/2014/09/2014-survey-monitoring-evaluation-v4.pdf.

Kremer, M., Gallant, S., Rostapshova, O., and Thomas, M. (2021) Is development economics a good investment? Evidence on scaling rate and social returns from USAID's innovation fund, Working Paper, https://cpb-us-w2.wpmucdn.com/voices.uchicago.edu/dist/0/2830/files/2021/09/SROR−21.03.12_clean−3.pdf.

Krippendorff, K. (2019) *Driving Innovation from Within*, New York: Columbia University Press.

Kruger, J. and Dunning, D. (1999) Unskilled and unaware of it: How difficulties in recognizing one's own incompetence lead to inflated self-assessments, *Journal of Personality and Social Psychology*, 77(6), 1121–1134, https://psycnet.apa.org/doiLanding?doi=10.1037%2F0022–3514.77.6.1121.

Kuchenbecker, K.J., Fiene, J., and Niemeyer, G. (2006) Improving contact realism through event-based haptic feedback, *IEEE Transactions on Visualization and Computer Graphics*, 12(2), 219–230, https://doi.org/10.1109/TVCG.2006.32.

Kumpf, B. and Jhunjhunwala, P. (2023) The adoption of innovation in international development organisations, OECD Development Co-operation Working Papers, No. 112, Paris: OECD Publishing, https://www.oecd.org/en/publications/the-adoption-of-innovation-in-international-development-organisations_21f63c69-en.html.

Kwet, M. (2019) Digital colonialism: US empire and the new imperialism in the Global South, *Race & Class*, 60(4), 3–26, https://journals.sagepub.com/doi/abs/10.1177/0306396818823172.

Ladcomm (2013) Universal Service Fund Study conducted on behalf of the GSM Association, London: GSMA, https://www.gsma.com/solutions-and-impact/connectivity-for-good/public-policy/wp-content/uploads/2016/09/GSMA2013_Report_SurveyOfUniversalServiceFunds.pdf.

Lane Fox, M. (2012) Letter to Francis Maude: DirectGov 2010 and beyond: Revolution not evolution, https://assets.publishing.service.gov.uk/media/5a78f39eed915d0422066a0d/Martha_20Lane_20Fox_s_20letter_20to_20Francis_20Maude_2014th_20Oct_202010.pdf.

Lenin, V.I. (1917) Империализм как высшая стадия капитализма, Petrograd (Pamphlet).

Lennie, J. and Tacchi, J. (2013) *Evaluating Communication for Development: A Framework for Social Change*, Abingdon: Earthscan.

Levendowski, A. (2018) How copyright law can fix Artificial Intelligence's implicit bias problem, *Washington Law Review*, 93(2), 579, https://digitalcommons.law.uw.edu/wlr/vol93/iss2/2/.

Lexxic (2024) *Levelling Up: Empowering Neurodiversity in the Tech Industry*, London: Lexxic, https://lexxic.com/blog/levelling-up-neuro-inclusion-in-the-tech-industry.

Lim, D. (2023) Generative AI and copyright: Principles, priorities and practicalities, *Journal of Intellectual Property Law and Practice*, 18(22), 841–842, https://academic.oup.com/jiplp/article/18/12/841/7331468.

Lipset, S.M. (1997) *American Exceptionalism*, New York: W.W. Norton.

Liyanage, H. (2009) *Sustainability First: In Search of Telecentre Sustainability*, Kotte: Sarvodaya Fusion, telecentre.org.

Lobosco, K. (2012) Doing good and making money, *Forbes*, https://www.forbes.com/sites/nyuentrepreneurschallenge/2012/10/15/doing-good-and-making-money/.

Locke, J. (1988) *Two Treatises of Government*, Laslett, P. (ed.), Cambridge: Cambridge University Press (originally published in 1689).

Low, P. (2025) Dr. Philip Low's post, https://www.linkedin.com/feed/update/urn:li:activity:7288439915485315072/.

Lowcock, M. and Dissanayake, R. (2024) *The Rise and Fall of the Department for International Development*, Washington, DC: Center for Global Development.

Ma, J., Lin, P., and Williams, J. (2023) Effectiveness of nature-based walking interventions in improving mental health in adults: A systematic review, *Current Psychology*, 43, 9521–9539, https://link.springer.com/article/10.1007/s12144-023-05112-z.

MacAskill, W. (2016) *Doing Good Better*, London: Guardian Faber.

Mackintosh, W. (2025) Foresighting viable open alternatives to address OER's existential threat from commercial generative AI: Confronting the unspoken challenge for Pacific small island developing states, OER Foundation, https://oerfoundation.org/wp-content/uploads/2025/01/OERF-Pacific-scenarios-for-GenAI–2025.pdf.

MacLean, D. (ed.) (2004) *Internet Governance: A Grand Collaboration*, New York: UN ICT Task Force, https://digitallibrary.un.org/record/552179.

Maclean, M., Harvey, C., Yang, R., and Mueller, F. (2021) Elite philanthropy in the United States and United Kingdom in the new age of inequalities, *International Journal of Management Reviews*, 23, 330–352, https://onlinelibrary.wiley.com/doi/10.1111/ijmr.12247.

Macon-Cooney, B. *et al.* (2024) *A New National Purpose: Leading the Biotech Revolution*, London: Tony Blair Institution for Global Change, https://assets.ctfassets.net/75ila1cntaeh/3pN0p29FVT0XV1FV9Up4MV/d2eaa86f106369827c3eed1b113c92ff/Tony_Blair_Institute_A_New_National_Purpose_-_Leading_the_Biotech_Revolution_January_2024.pdf.

Maddicott, J.R. (1975) The English peasantry and the demands of the Crown 1294–1341, *Past & Present, Supplement 1*.

Malaysia Government Portal (2025) My Government: High Speed Broadband, https://www.malaysia.gov.my/portal/content/30046.

Malkan, S. (2025) Critique of Gates Foundation's agricultural interventions in Africa, *U.S. Right to Know*, https://usrtk.org/bill-gates/critiques-of-gates-foundation/.

Manion, M. and Evan, W.M. (2000) The Y2K problem and professional responsibility: A retrospective analysis, *Technology in Society*, 22(3), 361–387, https://www.sciencedirect.com/science/article/abs/pii/S0160791X00000154.

Marcus, G. and Davis, E. (2019) *Rebooting AI: Building Artificial Intelligence We Can Trust*, New York: Pantheon.

Marker, P., McNamara, K., and Wallace, L. (2002) *The Significance of Information and Communication Technologies for Reducing Poverty*, London: DFID, https://www.ircwash.org/sites/default/files/Marker-2002-Significance.pdf.

Martens, J. (2007) *Multistakeholder Partnerships: Future Models of Multilateralism*? Berlin, Germany: Friedrich Ebert Stiftung, https://www.globalpolicy.org/sites/default/files/martens_multistakeholder_partnerships_online_version.pdf.

Martinsone, S. (2024) INGOs must confront their past – and help shape a different future, https://www.bond.org.uk/news/2024/10/ingos-must-confront-their-past-and-help-shape-a-different-future/.

Marx, K. (1844) A contribution to the critique of Hegel's Philosophy of Right, *Deutsch-Französische Jahrbücher*, 7 & 10 February 1844, Paris, https://www.marxists.org/archive/marx/works/1843/critique-hpr/intro.htm.

Masiero, S. (2023) ICT4D and critical data studies: A joint research agenda, in: Anderson, R., Vannini, S., Ahmed, S.I., Vashistha, A., and Raza, A.A. (eds) *ICTD '22: Proceedings of the 2022 International Conference on Information and Communication Technologies and Development*, Article 18, New York: Association for Computing Machinery, 1–5, https://dl.acm.org/doi/abs/10.1145/3572334.3572388.

Masiero, S. and Shakthi, S. (2020) Unique identification in India: Aadhaar, biometrics and technology-mediated identities, *South Asia Multidisciplinary Academic Journal*, 23, https://journals.openedition.org/samaj/6277.

Mattis, J.S., Hammond, W.P., Grayman, N., Bonacci, M., Brennan, W., Cowie, S-A., Ladyzhenskaya, L., and So, S. (2009) The social production of altruism: Motivations for caring action in a low-income urban community, *American Journal of Community Psychology*, 43(1–2), 71–84, https://pubmed.ncbi.nlm.nih.gov/19156513/.

Mawdsley, E. (2012) *From Recipients to Donors: Emerging Powers and the Changing Development Landscape*, London: Zed.

Mayer, H., Yee, L., Chui, M., and Roberts, R. (2025) *Superagency in the Workplace: Empowering People to Unlock AI'S Future Potential*, New York: McKinsey & Company, https://www.mckinsey.com/~/media/mckinsey/business-functions/quantumblack/our-insights/superagency-in-the-workplace-empowering-people-to-unlock-ais-full-potential-at-work/superagency-in-the-workplace-empowering-people-to-unlock-ais-full-potential-v3.pdf.

McCarthur, J.W. and Rasmusen, K. (2017) Change of pace: Accelerations and advances during the Millennium Development Goal era, Global Economy and Development Working Paper 98, Washington, DC: Brookings, https://www.brookings.edu/wp-content/uploads/2017/01/global_20170111_change_of_pace.pdf.

McKinsey (2020) Purpose: Shifting from why to how, https://www.mckinsey.com/capabilities/people-and-organizational-performance/our-insights/purpose-shifting-from-why-to-how.

McKinsey Global Institute (2018) *Smart Cities: Digital Solutions for a More Livable Future*, New York: McKinsey, https://www.mckinsey.com/~/media/mckinsey/industries/public-and-social-sector/our-insights/smart-cities-digital-solutions-for-a-more-livable-future/mgi-smart-cities-full-report.pdf.

MCMC (2024) Nadi will improve standard of living of local community – MCMC, https://mcmc.bernama.com/news.php?id=2315440.

MCMCTV (2013) Segalanya Mungkin dengan Jalur Lebar (Everything is possible with broadband), playlist of five videos, https://youtube.com/playlist?list=PLQVCnylUVwBZOGwLBZhibBhWmWCRv6aJz.

Meadows, D.H., Meadows, D.L., Randers, J., and Behrens, W.W. III (1972) *The Limits to Growth*, New York: Universe Books.

Megginson, W.L. (2005) *Privatization Industry Studies*, Oxford: Oxford University Press.

Meng, S-Q., Cheng, J-L., Li, Y-Y. *et al.* (2022) Global prevalence of digital addiction in general population: A systematic review and meta-analysis, *Clinical Psychology Review*, 92, 102128, https://pubmed.ncbi.nlm.nih.gov/35150965/.

Mental Health Foundation (2021) *Nature: How Connecting with Nature Helps our Mental Health*, London: Mental Health Foundation, https://www.mentalhealth.org.uk/our-work/research/nature-how-connecting-nature-benefits-our-mental-health.

Mettler, T. and Miscione, G. (2022) Is altruism dead? A critical case study on the paradigm shift in Open Government Data, *Information Polity*, 28(1), 199–218, https://journals.sagepub.com/doi/pdf/10.3233/IP−220048.

Mhlanga, D. (2023) *Fintech and Artificial Intelligence for Sustainable Development*, Cham: Palgrave Macmillan.

Millennium Villages Project (n.d.) About the Millennium Villages Project, https://www.un.org/esa/coordination/Alliance/Earth-Institute---The-Millennium-Villages-Project.htm.

Mills, J. (2024) How is digital accessibility awareness being promoted across the globe?, *GrackleDocs*, https://www.grackledocs.com/en/digital-accessibility-awareness/.

Milward-Oliver, G. (2005) *Maitland+20: Fixing the Missing Link*, Bradford on Avon: The Anima Centre.

Ministry of Human Resources Development, Government of India (2019) National Innovation and Startup Policy 2019 for students and faculty: A guiding framework for higher education students, New Delhi: Ministry of Human Resources Development, https://mic.gov.in/assets/doc/startup_policy_2019.pdf.

MIRI (n.d.) The problem, Machine Intelligence Research Institute, https://intelligence.org/the-problem/.

Mishra, V. (2025) UN faces 'race to bankruptcy' as Guterres unveils sharply reduced 2026 budget, *UN News*, https://news.un.org/en/story/2025/10/1166128.

Mitchell, J. and D'Rozario, J. (2022) Cost-effective EdTech Paper 2: Good practice, EdTech Hub, https://docs.edtechhub.org/lib/ZVX4DTXQ/download/RW5QDU64/Cost-Effective-EdTech-2-%E2%80%93-Good-Practice.pdf.

Moll, I. (2021) The myth of the Fourth Industrial Revolution, *Theoria*, 68(167), 1–38, https://www.berghahnjournals.com/view/journals/theoria/68/167/th6816701.xml.

Monino, J-L. and Sedkaoui, S. (2016) *Big Data, Open Data and Data Development*, London and Hoboken, NJ: ISTE and Wiley.

Monteiro, G., Moraes, L., Garcia, K., and Costa, V. da (2020) Auxílio emergencial: As histórias dos 'invisiveis' que buscam os R$ 600 para sobreviver, *O Globo Economia*, 1 May, https://oglobo.globo.com/economia/auxilio-emergencial-as-historias-dos-invisiveis-que-buscam-os-600-para-sobreviver−24405087.

Moore, D.A. and Healy, P.J. (2008) The trouble with overconfidence, *Psychological Review*, 115(2), 502–517, https://psycnet.apa.org/doiLanding?doi=10.1037%2F0033–295X.115.2.502.

Moran, C. (2020) On the rate of profit, *Rethinking Economics*, https://www.exploring-economics.org/en/discover/on-the-rate-of-profit/.

Moreno, M.R.G., Tzinova, K.T., Wang, Z., and Peršić, A. (2024) *Implementation of Science, Technology and Innovation policy instruments towards the Sustainable Development Goals*, Paris: UNESCO, https://sdgs.un.org/sites/default/files/2024-05/Moreno%2C-et-al._Implementation-of-Science%2C-Technology-and-Innovation-Policy-Instruments-towards-the-SDGs.pdf.

Morris, K. and Yoo, T. (2022) Oral history of Tae Yoo, Computer History Museum, https://archive.computerhistory.org/resources/access/text/2022/11/102792726-05-01-acc.pdf.

Mosse, D. (2011) *Adventures in Aidland: The Anthropology of Professionals in International Development*, London: Berghahn Books.

MRC (2012) *Good Research Practice: Principles and Guidelines*, London: MRC.

Munk, N. (2014) *The Idealist: Jeffrey Sachs and the Quest to End Poverty*, New York: Anchor.

Musk, E. (2025) Interview in *The Joe Rogan Experience*, 28 February, https://www.youtube.com/watch?v=sSOxPJD-VNo&t=4560s, at 1:16:00 mark.

MzN International (2023) How to create an agile, global, and digital-first non-profit organisation, Bond, https://www.bond.org.uk/news/2023/05/how-to-create-an-agile-global-and-digital-first-non-profit-organisation/.

Nabbale, H., Tumuhamye, N., Mugema, B., and Prigent, A. *(2023) Uganda's digital transformation journey, UNPD Uganda Blog,* https://www.undp.org/uganda/blog/ugandas-digital-transformation-journey.

Nelson, J., Hodges, A., Deri, C., Schneider, M., and Ruder, A. (2005) *Business and International Development: Opportunities, Responsibilities and Expectations A Survey of Global*

Opinion Leaders in Business, Civil Society and the Media, Cambridge, MA, London, and Chicago: Kennedy School of Governance, International Business Leaders Forum, and Edelman.
Newman, D., Jaciw, A.P., and Lazarev, V. (2018) *Guidelines for Conducting and Reporting EdTech Impact Research in US K-12 Schools*, Palo Alto, CA: Empirical Education and ETIN, https://www.empiricaleducation.com/pdfs/guidelines.pdf.
NHS (2018) Protocol, https://www.ukri.org/wp-content/uploads/2021/08/MRC-0208212-Good-research-practice_2014.pdf.
Niemitz, K. (2024) *Imperial Measurement: A Cost-Benefit Analysis of Western Colonialism*, London: Institute of Economic Affairs.
Nietzsche, F.W. (1980). *On the Advantage and Disadvantage of History for Life*, trans. Preuss, P., Indianapolis, IN: Hackett (originally published in German in 1874).
Nietzsche, F.W. (2008) *On the Genealogy of Morality: A Polemic*, Oxford: Oxford University Press (first published in German in 1887).
Nokia (n.d.) Towards universal access: Connecting rural communities via mobile communications, Nokia.
NormCoresky (2017) Overthrowing the Big Society: The Third Sector's neoliberal embrace, *New Socialist*, https://newsocialist.org.uk/overthrowing-the-big-society-the-third-sectors-neoliberal-embrace.
Nouwens, V. (2022) *A Transatlantic Approach to China in the Global Commons*, London: RUSI, https://static.rusi.org/322-WHR-Global-Commons.pdf.
Noxolo, P. (2016) Postcolonial approaches to development, in: Grugel, J. and Hammett, D. (eds), *Palgrave Handbook of International Development*, London: Palgrave Macmillan, 41–53.
Nwakanma, N. (2015) Nnenna Nwakanma on the Web in Africa, World Wide Web Foundation, https://webfoundation.org/2015/11/nnenna-nwakanma-on-the-web-in-africa/.
O'Boyle, E.J. (1999) Toward an improved definition of poverty, *Review of Social Economy*, 57(3), 281–301, https://www.tandfonline.com/doi/abs/10.1080/00346769900000003.
O'Connor, B., Cox, L., and Cooper, D. (2022) The ideology of American exceptionalism: American nationalism's nom de plume, *Journal of Political Ideologies*, 29(3), 634–655, https://www.tandfonline.com/doi/full/10.1080/13569317.2022.2112126.
OECD (2023) Human rights in the digital age, https://www.oecd.org/en/topics/sub-issues/human-rights-in-the-digital-age.html.
OECD (2025) *Tax Challenges Arising from the Digitalisation of the Economy – Consolidated Commentary to the Global Anti-Base Erosion Model Rules (2025): Inclusive Framework on BEPS*, Paris: OECD, https://www.oecd.org/en/publications/tax-challenges-arising-from-the-digitalisation-of-the-economy-consolidated-commentary-to-the-global-anti-base-erosion-model-rules-2025_a551b351-en.html.
OHCHR (1948) Universal Declaration of Human Rights, OHCHR, https://www.ohchr.org/en/human-rights/universal-declaration/translations/english.
Oliveira, A.L. (2022) A blueprint for conscious machines, *Proceedings of the National Academy of Sciences of the United States of America*, 119(23), e2205971119, https://www.pnas.org/doi/abs/10.1073/pnas.2205971119.
OLPC (2023) About us, https://laptop.org/aboutolpc/.
O'Neil, C. (2016) *Weapons of Math Destruction: How Big Data Increases Inequality and Threatens Democracy*, New York: Crown.
O'Neil, M. (2005) Remarks at the Telecentre.org launch, https://idl-bnc-idrc.dspacedirect.org/server/api/core/bitstreams/101865fb-eacb-47fa-9f9c-52acf732d4ba/content.
O'Neill, O. (1990) *Constructions of Reason: Explorations of Kant's Practical Philosophy*, Cambridge: Cambridge University Press.
O'Neill, O. (1996) *Towards Justice and Virtue*, Cambridge: Cambridge University Press.
O'Neill, O. (2016) *Justice Across Boundaries: Whose Obligations?*, Cambridge: Cambridge University Press.

Open Source Initiative (2006) History of the OSI, https://opensource.org/history.

Ord, T. (2014) Global poverty and the demands of morality, in: Perry, J. (ed.) *God, the Good, and Utilitarianism: Perspectives on Peter Singer*, Cambridge: Cambridge University Press, 177–191.

Ord, T. (2020) *The Precipice: Essential Risk and the Future of Humanity*, London: Bloomsbury Publishing.

O'Sullivan, S. (2025) *The Age of Diagnosis: Sickness, Health and Why Medicine Has Gone Too Far*, London: Hodder.

Oxfam (2024) Multilateralism in an era of global oligarchy, *Oxfam Media Briefing 23 September 2024*, Oxford: Oxfam, https://www.oxfam.org/en/research/multilaterialism-era-global-oligarchy.

Oxfam (2025) *Takers Not Makers: The Unjust Poverty and Unearned Wealth of Colonialism*, Oxford: Oxfam, https://www.oxfam.org/en/takers-not-makers-unjust-poverty-and-unearned-wealth-colonialism.

Panigrahi, S. (2021) Marginalized Aadhaar: Digital ID and exclusion of marginalized communities in India, Yoti, https://www.yoti.com/wp-content/uploads/Marginalized_Aadhaar-Subhashish-Panigrahi.pdf.

Park, M., Leahey, E., and Funk, R.J. (2023) Papers and patents are becoming less disruptive over time, *Nature*, 613, 138–144, https://www.nature.com/articles/s41586-022-05543-x.

Parry, R.L. (2024) Chinese army unveils robo wolf, the new leader of the AI pack, *The Times*, 12 November 2024, https://www.thetimes.com/world/asia/article/chinese-army-robot-wolf-artificial-intelligence-9l0zr6m58.

Parsheera, S. and Trehan, V. (2022) A structural analysis of the mobile telecommunications market: Exploring the Jio effect, in: Krishnaswamy, S. and Joshi, D. (eds), *The Philosophy and Law of Information Regulation in India*, Bengalaru: Centre for Law and Policy Research, https://publications.clpr.org.in/the-philosophy-and-law-of-information-regulation-in-india/chapter/a-structural-analysis-of-the-mobile-telecommunications-market-exploring-the-jio-effect/.

Patrick, S. (2023) The Universal Declaration of Human Rights at 75: An unfinished revolution, Carnegie Global Order and Institutions, https://carnegieendowment.org/posts/2023/12/the-universal-declaration-of-human-rights-at-75-an-unfinished-revolution?lang=en.

Patrick, S. (2025) The death of the world America made, *Emissary*, https://carnegieendowment.org/emissary/2025/02/trump-executive-order-treaties-organizations?lang=en.

Pauwels, L., Chalavi, S., and Swinnen, S.P. (2018) Aging and brain plasticity, *Aging*, 10(8), 1789–1790, https://pmc.ncbi.nlm.nih.gov/articles/PMC6128435/.

Perez, C. (2021) Foreign Policy analytics special reports: Global data governance, Parts 1 and 2, Washington, DC: Foreign Policy, https://foreignpolicy.com/2020/05/13/data-governance-privacy-internet-regulation-localization-global-technology-power-map/ and https://foreignpolicy.com/2020/06/26/data-governance-privacy-surveillance-covid-ai-encryption-technology-internet-power-map/.

Perry, H., Nur Adan, S., Malik, N., and Knight, S. (2025) *Advancing Digital Rights in 2025: Trends, Challenges, and Opportunities in the UK, EU and Global Landscape*, London: Demos, https://demos.co.uk/wp-content/uploads/2025/02/Digital-Rights-in–2025.ac_.pdf.

Piacentini, C., Bernardini, S., and Beck, J.C. (2019) Autonomous target search with multiple coordinated UAVs, *Journal of Artificial Intelligence Research*, 65, 11735, https://jair.org/index.php/jair/article/view/11635.

Pimienta, D. (2022) Resource: Indicators on the presence of languages in Internet, in: *Proceedings of the 1st Annual Meeting of the ELRA/ISCA Special Interest Group on Under-Resourced Languages*, Marseille, France, 83–91, https://aclanthology.org/2022.sigul–1.11/.

PitchDeckHunt (n.d.) Facebook's first pitch deck, https://www.pitchdeckhunt.com/pitch-decks/facebook.

Plous, S. (1993) *The Psychology of Judgment and Decision Making*, New York: McGraw-Hill.
Pogrebna, G. (2024) AI is a multi-billion dollar industry. It's underpinned by an invisible and exploited workforce, *The Conversation*, https://theconversation.com/ai-is-a-multi-billion-dollar-industry-its-underpinned-by-an-invisible-and-exploited-workforce−240568.
Pompa, C. (2013) Literature review on the impact of business incubation, mentoring, investment and training on start-up companies, London: Economics and Private Sector Professional Evidence and Applied Knowledge Services, for DFID, https://assets.publishing.service.gov.uk/media/57a08a21e5274a27b2000437/Literature_Review_on_the_Impact_of_Incubation_Investment_Training20.pdf.
Pope Francis (2013) *Evangelii Gaudium*, Rome: Vatican, https://www.vatican.va/content/francesco/en/apost_exhortations/documents/papa-francesco_esortazione-ap_20131124_evangelii-gaudium.html.
Potter, R.B. and Unwin, T. (1989) *The Geography of Urban-Rural Interactions in Developing Countries: Essays for Alan B. Mountjoy*, London: Routledge.
Prahalad, C.K. (2004) *The Fortune at the Bottom of the Pyramid: Eradicating Poverty Through Profits*, London: Financial Times and Prentice Hall.
Prahalad, C.K. and Hart, S.L. (2002) The fortune at the bottom of the pyramid, *Strategy + Business*, 26, https://www.strategy-business.com/article/11518.
Privacy International (2024) Humans in the AI loop: the data labelers behind some of the most powerful LLM's training datasets, https://privacyinternational.org/explainer/5357/humans-ai-loop-data-labelers-behind-some-most-powerful-llms-training-datasets.
Pun, M. (2023) *Mahabir Pun: Samjhana, Sapana Ra Abiral Yatra* (Mahabir Pun: Memories, Dreams, and Endless Journeys), Kathmandu: NIC.
Qi, J., Yu, L., Khoo, E.T. *et al.* (2024) Bridging the digital–physical divide using haptic and wearable technologies, *Nature Electronics*, 7, 1098–1110, https://www.nature.com/articles/s41928-024-01325-7.
Quiñones, D. and Rojas, L. (2023) Understanding the customer experience in human-computer interaction: A systematic literature review, *PeerJ Computer Science*, 10(9), https://peerj.com/articles/cs-1219/.
Rachlin, H. (2002) Altruism and selfishness, *Behavioral and Brain Science*, 25, 239–296, https://www.cambridge.org/core/journals/behavioral-and-brain-sciences/article/abs/altruism-and-selfishness/B8957ED895AB2E66183F128433F6C9E9.
Ramadhan, R.N., Rampengan, D.D., Yumnanisha, D.A. *et al.* (2024) Impacts of digital social media detox for mental health: A systematic review and meta-analysis, *Narra Journal*, 4(2), e786, https://pmc.ncbi.nlm.nih.gov/articles/PMC11392003/.
Ramalingam, B. and Bound, K. (eds) (2016) *Innovation for International Development: Navigating the Paths and Pitfalls*, London: Nesta.
Ramalingam, B., Hernandez, K., Martin, P., and Faith, B. (2016) *Ten Frontier Technologies for International Development*, Brighton: IDS, https://opendocs.ids.ac.uk/articles/online_resource/Ten_Frontier_Technologies_for_International_Development/26440834.
Rand, A. (1964) *The Virtue of Selfishness: A New Concept of Egoism*, London: Signet.
Rawls, J. (1971) *A Theory of Justice*, Cambridge, MA: Harvard University Press.
Rawls, J. (2001) *Justice as Fairness: A Restatement*, Cambridge, MA: Harvard University Press.
Razzall, K. (2025) Bill Gates: We've given away $100bn, but my children won't be poor when I'm gone, BBC, https://www.bbc.co.uk/news/articles/cx2p4p4l78zo.
Reginster, B. (2000) Nietzsche on selflessness and the value of altruism, *History of Philosophy Quarterly*, 17(2), 177–200, https://www.jstor.org/stable/27744848.
Reidy, S. (2019) AGCO investing in agri-industry park in Mozambique, *World Grain.com*, https://www.world-grain.com/articles/12235-agco-investing-in-agri-industry-park-in-mozambique.
Renehan, E.J. (2006) *Dark Genius of Wall Street: The Misunderstood Life of Jay Gould, King of the Robber Barons*, New York: Basic Books.

Riding, J.B. (2022) An evaluation of the process of peer review, *Palynology*, 47(1), 2151052, https://www.tandfonline.com/doi/permissions/10.1080/01916122.2022.2151052.
Rigg, S. (2024) The Summit of the Future: A bold success of a tepid failure, Global Network of Civil Society Organisations for Disaster Reduction, https://www.gndr.org/summit-of-the-future-success-or-tepid-failure.
Rising Academies (2025) LearnLens, https://www.risingacademies.com/learnlens.
Roberts, E. (n.d.) GNU, https://cs.stanford.edu/people/eroberts/courses/cs181/projects/2000-01/open-source/gnu.htm.
Roberts, T. (2025) *Understanding Digital Rights: Definitions, Conceptions, and Myths*, Brighton: Institute of Development Studies, https://www.ids.ac.uk/publications/understanding-digital-rights-definitions-conceptions-and-myths/.
Rodriguez, A.M. and Woodhouse, T. (2022) Mobile data costs have increased, making internet connectivity unaffordable for many, *Alliance for Affordable Internet*, https://a4ai.org/news/mobile-data-cost-have-increased-making-internet-connectivity-unaffordable-for-many/.
Rodriguez-Segura, D. (2021) EdTech in developing countries: A review of the evidence, *World Bank Research Observer*, 37(2), 171–203, https://academic.oup.com/wbro/article-abstract/37/2/171/6333790.
Rodrik, D. (2017) Rescuing economics from neoliberalism, *The Boston Review*, https://www.bostonreview.net/articles/dani-rodrik-rescuing-economics-neoliberalism.
Rogers, E.M. (1962) *Diffusion of Innovations*, New York: Free Press of Glencoe.
Rogers, Y. (2012) *HCI Theory: Classical, Modern and Contemporary*, Cham: Springer Nature.
Rogerson, S. and Rogerson, A. (2007) Digital slavery, *IMIS Journal*, 17(5), https://www.researchgate.net/publication/313619199_Digital_Slavery.
Rosling, H., Rünnlund, A.R., and Rosling, O. (2018) *Factfulness: Ten Reasons We're Wrong about the World – And Why Things Are Better Than You Think*, London: Sceptre.
Rostow, W.W. (1960) *The Stages of Economic Growth: A Non-Communist Manifesto*, Cambridge: Cambridge University Press.
Roy, A. (2024) How India is using the Internet, *Economic Times*, 10 March, https://economictimes.indiatimes.com/tech/technology/how-india-is-using-the-internet/articleshow/108354854.cms.
Ruiters, T-L. (2023) Spotlight on Parklands sex trade, *Weekend Argus*, https://iol.co.za/weekend-argus/news/2022-05-22-spotlight-on-parklands-sex-trade/.
Rushton, S. and Williams, O.D. (eds) (2011) *Partnerships and Foundations in Global Health Governance*, Basingstoke: Palgrave Macmillan.
Rustad, S.A. (2024) *Conflict Trends: A Global Overview*, Oslo: Peace Research Institute, PRIO Paper 2024.
Sachs, J. (2005) *The End of Poverty: How We Can Make It Happen in Our Lifetime*, London: Penguin Books.
Sachs, J. (2008) The digital war on poverty, *Project Syndicate*, 20 August, https://www.project-syndicate.org/commentary/the-digital-war-on-poverty.
Sachs, J. (2018) Lessons from the Millennium Villages project: A personal perspective, *The Lancet Global Health*, 6(5), E472–E474, https://www.thelancet.com/journals/langlo/article/PIIS2214–109X(18)30199–2/fulltext.
Sadhya, D. and Sahu, T. (2024) A critical survey of the security and privacy aspects of the Aadhaar framework, *Computers and Security*, 140, 103782, https://www.sciencedirect.com/science/article/abs/pii/S016740482400083X.
Sadowsky, G. (ed.) (2012) *Accelerating Development Using the Web: Empowering Poor and Marginalised Populations*, World Wide Web Foundation.
Sandel, M. (1998) *Liberalism and the Limits of Justice*, New York: Cambridge University Press.
Santos, R.M.S., Mendes, C.G., Bressani, G.Y.S. *et al.* (2023) The associations between screen time and mental health in adolescents: A systematic review, *BMC Psychology*, 11, 127, https://bmcpsychology.biomedcentral.com/articles/10.1186/s40359-023-01166-7.

Saraswati, J. (2014) Konza City and the Kenyan software services strategy: The great leap backward?, *Review of African Political Economy*, 41(S1), S128–S137, https://www.jstor.org/stable/24858302.

Schelenz, L. and Pawelec, M. (2022) Information and Communication Technologies for Development (ICT4D) critique, *Information Technology for Development*, 28(1),165–188, https://www.tandfonline.com/doi/full/10.1080/02681102.2021.1937473.

Schwab, K. (2017) *The Fourth Industrial Revolution*, London: Penguin.

Schwab, T. (2023) *The Bill Gates Problem: Reckoning with the Myth of the Good Billionaire*, London: Penguin Business.

Schwab, T. (2025) What is Bill Gates afraid of? The mask is finally coming off, *UnHerd*, 4 February, https://unherd.com/2025/02/what-is-bill-gates-afraid-of/.

Scullin, M.H. (2022) Walt Kelly – Pogo – 19 May 1968 – historic first appearance of a personal computer in print, https://www.comicartfans.com/gallerypiece.asp?piece=1794160.

Sein, M.K., Thapa, D., Hatakka, M., and Sæbø, Ø. (2019) A holistic perspective on the theoretical foundations for ICT4D research, *Information Technology for Development*, 25(1), 7–25, https://www.tandfonline.com/doi/full/10.1080/02681102.2018.1503589/

Selinger, M. (2001) The Imfundo Project: ICT in teacher education in developing countries, in: Price, J., Willis, D., Davis, N. and Willis, J. (eds), *Proceedings of SITE 200—Society for Information Technology & Teacher Education International Conference*, 3008–3013, Norfolk, VA: Association for the Advancement of Computing in Education (AACE), https://link.springer.com/chapter/10.1007/978-0-387-35596-2_80.

Sen, A. (1992) *Inequality Reexamined*, Cambridge, MA: Harvard University Press.

Sen, A. (1999) *Development as Freedom*, New York: Alfred Knopf.

Sen, A. (2010) *The Idea of Justice*, London: Penguin.

Servon, L.J. (2002) *Bridging the Digital Divide: Technology, Community and Public Policy*, Oxford: Blackwell.

Seth, A. (2022) *Technology and (Dis)Empowerment: A Call to Technologists*, Bingley: Emerald.

Shafqat, M. (2016) Strategic project: Incubation centre to foster growth, says Anusha, *The Express Tribune*, 19 July, https://tribune.com.pk/story/1145216/strategic-project-incubation-centre-foster-growth-says-anusha.

Sharma, A., Pandey, S., and Sethi, A. (2024) *The Making of the Greatest. 4 Books Collection (Jack Ma, Mark Zuckerberg, Jeff Bezos, Bill Gates)*, Delhi: Rupa Publications.

Sherif, A. (2024) Global market share of the information and communication technology (ICT) market from 2013 to 2024, by selected country, *Statista*, https://www.statista.com/statistics/263801/global-market-share-held-by-selected-countries-in-the-ict-market.

Sholl, J. (2017) The muddle of medicalization: Pathologizing or medicalizing?, *Theoretical Medicine and Bioethics*, 38, 265–278, https://link.springer.com/article/10.1007/s11017-017-9414-z.

Siele, M.K.N. (2023) M-Pesa has been huge for Kenya's economy – and for scammers, *Rest of World*, https://restofworld.org/2023/mpesa-sim-swap-fraud/.

Sikolia, E. (2025) The Kenyan Historian: I visited Konza – the smart city Kenya is constructing to see the progress, https://www.youtube.com/watch?v=LYrMXd6S-uM.

Simanis, E. (2012) Reality check at the bottom of the pyramid, *Harvard Business Review*, June, https://hbr.org/2012/06/reality-check-at-the-bottom-of-the-pyramid.

Simmons and Simmons (2022) In brief: Telecoms regulation in European Union, *Lexology*, https://www.lexology.com/library/detail.aspx?g=b1877b9d-a036–475e-a573-f1eb0211cd8a.

Singer, P. (1979) *Practical Ethics*, Cambridge: Cambridge University Press.

Singer, P. (2002) *One World: The Ethics of Globalisation*, New Haven, CT: Yale University Press.

Singer, P. (2015) *The Most Good You Can do: How Effective Altruism Is Changing Ideas About Living Ethically*, New Haven, CT: Yale University Press.

Singh, A., Doval, P., and Kurian, B. (2016) I wish I were 30 years old today: Mukesh Ambani, *The Times of India*, 8 September, https://timesofindia.indiatimes.com/business/india-business/i-wish-i-were-30-years-old-today-mukesh-ambani/articleshow/54122922.cms.

Singh, P. and Jaeckel, A. (2024) Ownership in the deep seas. Undermining by mining? Deep seabed mining in light of international marine environmental law, *AJIL Unbound*, 8, 72–77, https://doi.org/10.1017/aju.2024.8

Skyquest (2025) Smartphones market size, share, and growth analysis, https://www.skyquestt.com/report/smartphones-market.

Smart Africa (2017) *Smart Sustainable Cities: A Blueprint for Africa*, Kigali: Smart Africa.

Smart-Villages (2023) Empowering African communities through sustainable development, https://smart-villages.org/.

Smart Villages Research Group (2017) Smart Villages: A pocket guide to rural energy and 'smart villages', Abingdon: Smart Villages Research Group, https://e4sv.org/wp-content/uploads/2017/06/PocketGuideweb.pdf.

Smith, N. (2025) The dawn of the posthuman age, *Noahpinion*, https://www.noahpinion.blog/p/the-dawn-of-the-posthuman-age.

Sneyd, A. (2024) *Hidden Politics in the UN Sustainable Development Goals*. Rugby: Practical Action.

Sordi, D. de (2023) Empobrecimento, fome e pandemia: O Auxílio Emergencial, o fim do Programa Bolsa Família e oi Auxílio Brasil, 2019–2022, *História, Ciências, Saude-Manguinhos*, 30 (suppl. 1), 1–20, https://www.scielo.br/j/hcsm/a/Zz97SwmZnfrqSz4GSG6Q6Rx/?format=html&lang=pt.

Souter, D. (2007) Internet governance and development: another digital divide?, *Information Polity*, 12(1–2), https://journals.sagepub.com/doi/10.3233/IP-2007-0111.

Souter, D. (2024) Inside the Digital Society: Good (digital) governance, *APC*, https://www.apc.org/en/blog/inside-digital-society-good-digital-government.

Spar, D. (2015) *Pirates, Prophets and Pioneers: Business and Politics Along the Technological Frontier*, New York: Random House.

Spies Coaching (2023) How Jeff Bezos' ADHD helped him become a successful entrepreneur, https://www.spiescoaching.co.uk/blog/how-jeff-bezos-adhd-helped-him-become-a-successful-entrepreneur.

Spitzer, M. (2014) *Digitale Demenz: Wie wir uns und unsere Kinder um den Verstand bringen*, Munich: Droemer.

Stang, G. (2013) Global Commons: Between cooperation and competition, European Union Institute for Strategic Studies, Brief 17, https://www.iss.europa.eu/publications/briefs/global-commons-between-cooperation-and-competition.

Statcounter (2025) Operating System market share worldwide, https://gs.statcounter.com/os-market-share.

Stiglitz, J.E. (2006) *Making Globalization Work*, New York: W.W. Norton & Company.

Stiglitz, J.E. (2012) *The Price of Inequality: How Today's Divided Society Endangers Our Future*, London: Allen Lane.

Stiglitz, J.E. (2016) *The Great Divide: Unequal Societies and What We Can Do About Them*, New York: W.W. Norton & Company.

Strubell, E., Ganesh, A., and McCallum, A. (2020) Energy and policy considerations for modern deep learning research, *Proceedings of the AAAI Conference on Artificial Intelligence*, 34(09), 13693–13696, https://ojs.aaai.org/index.php/AAAI/article/view/7123.

Swaminathan, M.S. and Kesavan, P.C. (2016) Achieving the Sustainable Development Goals, *Current Science*, 110(2), 127–128, https://www.jstor.org/stable/24906722.

Teleanu, S. (2024) *Unpacking Global Digital Compact: Actors, Issues and Processes*, Msida: Diplo.

telecentre.org (2025) Home page, https://www.telecentre.org/.

Tennant, J.P. and Ross-Hellauer, T. (2020) The limitations to our understanding of peer review, *Research Integrity and Peer Review*, 5, 6, https://researchintegrityjournal.biomedcentral.com/articles/10.1186/s41073-020-00092-1.

Tennyson, R. (2003) *The Partnering Toolbook*, London: IBLF and GAIN.

Terbeche, M. and Carrier, M. (2019) An 'agile manifesto' for humanitarian and development projects, *Groupe URD*, https://www.urd.org/en/review-hem/an-agile-manifesto-for-humanitarian-and-development-projects/.

Thapa, D. (2025) Ethics-based ontology in ICT4D, *Information Technology for Development*, https://doi.org/10.1080/02681102.2025.2498925.

The Bee (2005) Heard in the hive, *Times of Malta*, 26 November, https://timesofmalta.com/article/heard-in-the-hive.70729.

The Digital School (2023) Driving the frontiers of digital education: The Digital School's progress and impact report, Dubai: The Digital School, https://thedigitalschool.org/publications/driving-the-frontiers-of-digital-education/.

The Digital School (2025) About The Digital School, https://thedigitalschool.org/about-us/.

The New Times (2008) Top Mineduc official arrested over "graft", *The New Times*, 16 November, https://www.newtimes.co.rw/article/13485/top-mineduc-official-arrested-over-agrafta.

Thomson, A. (2025) Neurodiversity is a power we must tap into, *The Times*, 28 January, https://www.thetimes.com/comment/columnists/article/neurodiversity-is-a-power-we-must-tap-into-2t56jj8pr.

Thorner, D., Kerblay, B., and Smith, R.E.F. (eds) (1966) *A.V. Chayanov on The Theory of Peasant Economy*, Homewood, IL: The American Economic Association.

Thorpe, J., Guijt, J., Sprenger, T., and Stibbe, D. (2021) *Multi Stakeholder Platforms as System Change Agents: A Guide for Assessing Effectiveness*, Wageningen: Wageningen Centre for Development Innovation, https://research.wur.nl/en/publications/multi-stakeholder-platforms-as-system-change-agents-a-guide-for-a/.

Tilton, Z., Harnar, M., Raftree, L., Perrin, P., Bruening, G., Banerji, S., Gordley, J., McGuigan, M., Foster, H., and Behr, M. (2020) What we know about traditional MERL Tech: Insights from a scoping review, https://merltech.org/wp-content/uploads/2020/07/2_MERL_Tech-Insights_FINAL_7.16.2020.pdf.

Topping, J., Jenks, B., Hammargren, H., Knutsson, P. *et al.* (2021) *Financing the UN Development System: Time to Meet the Moment*, Uppsala: Dag Hammarskjöld Foundation and UN Multi-Partner Trust Fund Office, https://www.daghammarskjold.se/wp-content/uploads/2021/08/dhf-time-to-meet-the-moment-executive-summary-2021-final.pdf.

Toyama, K. (2015) *Geek Heresy: Rescuing Social Change from the Cult of Technology*, New York: Public Affairs.

Trepte, S. and Masur, P.K. (eds) (2023) *The Routledge Handbook of Privacy and Social Media*, London: Routledge.

Trucano, M. (2005) *Knowledge Maps: ICTs in Education*, Washington, DC: *info*Dev, World Bank, https://documents1.worldbank.org/curated/en/457411468341334749/pdf/319530WP0REVIS0800PUBLIC0Box379827B.pdf.

Tsvetkova, M., Yasseri, T., Pescetelli, N., and Werner, T. (2024) A new sociology of humans and machines, *Nature Human Behaviour*, 8, 1864–1876.

UC Davis (2023) Feed the Future Innovation Lab for Horticulture: History, https://horticulture.ucdavis.edu/history.

Uganda Communications Commission (UCC) (2024) Market Performance Report – Q4 2024 (Oct.–Dec. 2024), https://www.ucc.co.ug/wp-content/uploads/2025/02/UCC-Market-Report-for-Q4–2024-December−2024.pdf.

UK (2018) Rules Based International System Conference, *Gov.UK*, https://www.gov.uk/government/news/rules-based-international-system-conference.

UK Addiction Treatment Centres (2025) Addiction in the digital age: The new epidemics, https://www.ukat.co.uk/blog/society/addiction-in-the-digital-age-the-new-epidemics/.
UKRI (2019) *UK Research and Innovation: Annual Report and Accounts 2018–2019*, Swindon: UKRI.
UN (2004) *Beyond Conventional Wisdom in Development Policy: and Intellectual History of UNCTAD 1964–2004*, New York: UN.
UN (2010) We can end poverty: 2015 Millennium Goals, https://www.un.org/millenniumgoals/pdf/MDG_FS_1_EN.pdf.
UN (2016) Final list of proposed Sustainable Development Goal indicators, UN, https://sustainabledevelopment.un.org/content/documents/11803Official-List-of-Proposed-SDG-Indicators.pdf.
UN (2018) UN Secretary-General's Strategy on New Technologies, https://www.un.org/en/newtechnologies/images/pdf/SGs-Strategy-on-New-Technologies.pdf.
UN (2020) The UN Secretary-General's Roadmap on Digital Cooperation, https://www.un.org/en/sg-digital-cooperation-panel.
UN (2021a) *Our Common Agenda: Report of the Secretary-General*, New York: UN, https://www.un.org/en/content/common-agenda-report/assets/pdf/Common_Agenda_Report_English.pdf.
UN (2021b) Mr. Fabrizio Hochschild Drummond of Chile – Secretary-General's Envoy on Technology, UN Secretary-General Personnel Appointments, 22 January 2021, https://www.un.org/sg/en/content/sg/personnel-appointments/2021-01-22/mr-fabrizio-hochschild-drummond-of-chile-secretary-generals-envoy-technology.
UNGA (2025) Revised estimates relating to the proposed programme budget for 2026 and the support account for peacekeeping operations for the 2025/26 period, Report of the Secretary General (UNGA A/80/400), https://docs.un.org/en/A/80/400.
UN (2023a) Members of the High-Level Advisory Body on Artificial Intelligence, https://www.un.org/en/ai-advisory-body/members.
UN (2023b) National commitments to SDG transformation, https://sdgs.un.org/SDGSummitActions/National.
UN (2024a) *Governing AI for Humanity*, New York: UN, https://digitallibrary.un.org/record/4062495.
UN (2024b) *Summit of the Future Outcome Documents September 2024: Pact for the Future, Global Digital Compact and Declaration on Future Generations*, New York: UN, https://www.un.org/sites/un2.un.org/files/sotf-pact_for_the_future_adopted.pdf.
UN (2024c) *The Sustainable Development Goals Report 2024*, New York: UN, https://unstats.un.org/sdgs/report/2024/The-Sustainable-Development-Goals-Report–2024.pdf.
UN (2024d) The UN system, https://www.un.org/en/about-us/un-system.
UN (2024e) Voluntary endorsement of the Global Digital Compact, https://www.un.org/global-digital-compact/sites/default/files/2024-10/voluntary_endorsement_of_the_global_digital_compact_-_explanatory_note_0.pdf.
UN (n.d.) United Nations Charter (full text), https://www.un.org/en/about-us/un-charter/full-text.
UNDESA (2015) *Transforming Our World: The 2030 Agenda for Sustainable Development*, New York: UNDESA, https://sdgs.un.org/2030agenda.
UNDESA (2018) *World Economic and Social Survey 2018: Frontier Technologies for Sustainable Development*, New York: UNDESA, https://desapublications.un.org/publications/world-economic-and-social-survey-2018-frontier-technologies-sustainable-development.
UNDESA (2024) *UN E-Government Survey 2024*, New York: UNDESA, https://desapublications.un.org/publications/un-e-government-survey–2024.
UNDP (2022) How UNDP and ITU are strengthening government capacity for digital transformation, UNDP, https://www.undp.org/digital/stories/how-undp-and-itu-are-strengthening-government-capacity-digital-transformation.

UNECE (2020) Measuring and monitoring progress towards the Sustainable Development Goals. Geneva: UNECE, https://unece.org/unece-and-sdgs/publications/measuring-and-monitoring-progress-towards-sustainable-development-goals.

UNEP (2024) Artificial Intelligence (AI) end-to-end: The environmental impact of the full AI lifecycle needs to be comprehensively assessed, issue note, https://wedocs.unep.org/20.500.11822/46288.

UNESCO (2022a) *The 2019 UNESCO Recommendation on Open Educational Resources (OER): Supporting Universal Access to Information Through Quality Open Learning Materials*, Paris: UNESCO, https://unesdoc.unesco.org/ark:/48223/pf0000383205.

UNESCO (2022b) *An Introduction to the UNESCO Recommendation on Open Science*, Paris: UNESCO, https://www.unesco.org/en/articles/introduction-unesco-recommendation-open-science.

UNESCO (2022c) *Recommendation on the Ethics of Artificial Intelligence Adopted on 23 November 2021*, Paris: UNESCO, https://www.unesco.org/en/articles/recommendation-ethics-artificial-intelligence.

UNESCO (2023) *Open Science Toolkit*, Paris: UNESCO, https://www.unesco.org/en/open-science/toolkit.

UNESCO (2025) Aid to education: Time for tough decisions, Policy Paper 50, June 2025, https://unesdoc.unesco.org/ark:/48223/pf0000394153.

UNESCO (n.d.) Ethics of Artificial Intelligence: The recommendation, https://www.unesco.org/en/artificial-intelligence/recommendation-ethics.

UNGA (2000) Resolution adopted by the General Assembly, 55/2, United Nations Millennium Declaration, A/Res/55/2, New York: UN, https://www.un.org/en/development/desa/population/migration/generalassembly/docs/globalcompact/A_RES_55_2.pdf.

UNGA (2011) Report of the Special Rapporteur on the promotion and protection of the right to freedom of opinion and expression, Frank La Rue, UNGA Human Rights Council, 17th Session, Agenda Item 3, A/HRC/17/27, https://www2.ohchr.org/english/bodies/hrcouncil/docs/17session/A.HRC.17.27_en.pdf.

UNGA (2024) Revised estimates resulting from General Assembly resolution 79/1, entitled "The Pact for the Future", Report of the Secretary-General (UNGA A/79/583), https://documents.un.org/doc/undoc/gen/n24/338/32/pdf/n2433832.pdf.

UNGA (2025) Revised estimates relating to the proposed programme budget for 2026 and the support account for peacekeeping operations for the 2025/26 period, Report of the Secretary-General (UNGA A/80/400), https://docs.un.org/en/A/80/400.

UN GDC (2025) Endorsements, 14 March 2025, https://www.un.org/global-digital-compact/en/endorsements.

UNICEF (2014) Learning from failures: Part 1, https://www.unicef.org/innovation/stories/learning-failures-part−1.

UNIN (2023) What is the UN Innovation Network?, https://www.uninnovation.network/who-is-the-unin.

UNODET (n.d.) High-Level Advisory Body on Artificial Intelligence, https://www.un.org/digital-emerging-technologies/ai-advisory-body.

UNOSET (2022) Achieving universal and meaningful digital connectivity: Setting a baseline and targets for 2030. New York: UNOSET, https://www.itu.int/itu-d/meetings/statistics/wp-content/uploads/sites/8/2022/04/UniversalMeaningfulDigitalConnectivityTargets2030_BackgroundPaper.pdf.

UNOSET (2024) Global Digital Compact. New York: UNOSET. https://www.un.org/global-digital-compact/sites/default/files/2024-09/Global-Digital-Compact---English_0.pdf.

UNSD (2023) Progress chart. New York: UN Statistical Division https://unstats.un.org/sdgs/report/2023/progress-chart/.

UNSG HLP (2019) The age of digital interdependence. Report of the UN Secretary-General's High-level Panel on Digital Cooperation, June 2019, https://www.un.org/en/pdfs/HLP-on-Digital-Cooperation-Report-Executive-Summary---ENG.pdf.

Unwin, H.J.T., Cori, A., Imai, N., Gaythorpe, K.A.M., Bhatia, S., Cattarino, L., Donnelly, C.A., Ferguson, N.M. and Baguelin, M. (2022) Using next generation matrices to estimate the proportion of infections that are not detected in an outbreak, *Epidemics*, 41, 100637, https://www.sciencedirect.com/science/article/pii/S1755436522000779.

Unwin, H.J.T., Sherrard-Smith, E., Churcher, T.S. and Ghani, A.C. (2023) Quantifying the direct and indirect protection provided by insecticide-treated bed nets against malaria, *Nature Communications*, 14, 676, https://www.nature.com/articles/s41467-023-36356-9.

Unwin, T. (1987) Household characteristics and agrarian innovation adoption in North-West Portugal, *Transactions of the Institute of British Geographers*, 12(2), 131–146.

Unwin, T. (1992) *The Place of Geography*, Harlow: Longman.

Unwin, T. (ed.) (1994) *Atlas of World Development*, Chichester: Wiley.

Unwin, T. (2004) ICT and education in Africa: Partnership, practice and knowledge sharing, *Review of African Political Economy*, 31(9), 150–160, https://www.jstor.org/stable/4006949?seq=1.

Unwin, T. (2005a) Towards a framework for the use of ICT in teacher training in Africa, *Open Learning: The Journal of Open and Distance Education*, 20(2), 113–129, https://www.tandfonline.com/doi/full/10.1080/02680510500094124.

Unwin, T. (2005b) *Partnerships in Development Practice. Evidence from Multi-Stakeholder ICT4D Partnership Practice in Africa*, Paris: UNESCO for the World Summit on the Information Society, https://unesdoc.unesco.org/ark:/48223/pf0000142982.

Unwin, T. (2007) No end to poverty, *Journal of Development Studies*, 45(3), 929–953, https://www.tandfonline.com/doi/abs/10.1080/00220380701384596.

Unwin, T. (2008) On the richness of Africa, https://timunwin.blog/wp-content/uploads/2024/11/on-the-richness-of-africa-small.pdf.

Unwin, T. (ed.) (2009) *ICT4D: Information and Communication Technology for Development*, Cambridge: Cambridge University Press.

Unwin, T. (2011) On publishing in ICT4D, https://timunwin.blog/2011/01/02/on-publishing-in-ict4d/.

Unwin, T. (2014) Prolegomena on Human Rights and Responsibilities, https://timunwin.blog/2014/09/01/prolegomena-on-human-rights-and-responsibilities/.

Unwin, T. (2015a) MultiStakeholder partnerships in Information and Communication for Development Interventions, in: Mansell, R. and Ang, P.H. *et al.* (eds), *International Encyclopedia of Digital Communication and Society*, Chichester: Wiley, 634–644, https://onlinelibrary.wiley.com/doi/abs/10.1002/9781118767771.wbiedcs047.

Unwin, T. (2015b) ICTs and the failure of the Sustainable Development Goals, https://timunwin.blog/2015/08/05/icts-and-the-failure-of-the-sustainable-development-goals/.

Unwin, T. (2017a) *Reclaiming Information and Communication Technologies for Development*, Oxford: Oxford University Press.

Unwin, T. (2017b) ICTs, citizens and the state: Moral philosophy and development practices, *The Electronic Journal of Information Systems in Developing Countries*, 44(1), 1–16, https://onlinelibrary.wiley.com/doi/10.1002/j.1681-4835.2010.tb00313.x.

Unwin, T. (2018) ICTs and the failure of the SDGs, https://timunwin.blog/2018/04/23/icts-and-the-failure-of-the-sdgs/.

Unwin, T. (2019) Why the notion of a Fourth Industrial Revolution is so problematic, https://timunwin.blog/2019/03/09/why-the-notion-of-a-fourth-industrial-revolution-is-so-problematic/.

Unwin, T. (2020a) Guidance note 12: Ensuring rigorous monitoring and evaluation of initiatives using digital technologies in education for the most marginalised, in: Unwin, T., Naseem, A., Pawluczuk, A., Shareef, M., Spiesberger, P., West, P. and Yoo, C. (eds), *Education for the Most Marginalised Post-COVID-19: Guidance for Governments on the Use of Digital Technologies in Education*. London: UNESCO Chair in ICT4D and EdTech Hub, https://edtechhub.org/wp-content/uploads/2020/09/Education-for-the-most-marginalised-Report-Guidance-note-12-v8.pdf.

Unwin, T. (2020b) Digital technologies and climate change, Part III: Policy implications towards a holistic appraisal of digital technology sector, https://timunwin.blog/2020/01/16/digital-technologies-and-climate-change-part-iii-policy-implications-towards-a-holistic-appraisal-of-digital-technology-sector/.

Unwin, T. (2021) A new UN for a new (and better) global order (Part One): Seven challenges, https://timunwin.blog/2021/07/02/a-new-un-for-a-new-and-better-global-order-part-one/.

Unwin, T. (2022a) On language, gender and digital technologies, in: Fiormonte, D., Ricaurte, P. and Chaudhuri, S. (eds), *Global Debates in the Digital Humanities*, Minneapolis, MN: University of Minnesota Press, 298–304.

Unwin, T. (2022b) Freedom, enslavement and the digital barons: A thought experiment, https://timunwin.blog/2022/05/25/freedom-enslavement-and-the-digital-barons-a-thought-experiment/.

Unwin, T. (2023a) An environmentally harmful alliance of growth mantras, https://timunwin.blog/2023/05/17/an-environmentally-harmful-alliance-of-growth-mantras/.

Unwin, T. (2023b) Reflections on the Global Digital Compact, https://timunwin.blog/2023/04/20/reflections-on-the-global-digital-compact/.

Unwin, T. (2023c) Reflections on slavery: Past, present and future, https://timunwin.blog/2023/03/08/reflections-on-slavery-past-present-and-future/.

Unwin, T. (2024a) Scientism, multistakeholderism and the Global Digital Compact, https://timunwin.blog/2024/07/09/scientism-multistakeholderism-and-the-global-digital-compact/.

Unwin, T. (2024b) Why "we" (the people of the world) need to reject the Global Digital Compact, https://timunwin.blog/2024/06/11/why-we-the-people-of-the-world-need-to-reject-the-global-digital-compact/.

Unwin, T., Casentini, G., Harindranath, G. and Lorini, M.R. (2023) What works for migrants: Reflections on research practice in the interests of migrants, Egham: ICT4D Collective, Working Paper 1 (v.7), https://ict4d.org.uk/wp-content/uploads/2023/11/wp-1-v7.pdf.

Unwin, T., Naseem, A., Pawluczuk, A., Shareef, M., Spiesberger, P., West, P., and Yoo, C. (2020) *Education for the Most Marginalised post-COVID-19: Guidance for Governments on the Use of Digital Technologies in Education.* London: UNESCO Chair in ICT4D and EdTech Hub, https://docs.edtechhub.org/lib/JDSSXW7Q.

Unwin, T., Tan, M., and Pauso, K. (2017) The potential of e-Learning to address the needs of out-of-school youth in the Philippines, *Children's Geographies*, 5(4), 443–462, https://www.tandfonline.com/doi/full/10.1080/14733280701631940.

Unwin, T. and Wong, A. (2012) *Global Education Initiative: Retrospective on Partnerships for Education Development 2003–2011*, Geneva: World Economic Forum, https://www.weforum.org/publications/global-education-initiative-retrospective-partnerships-education-development-2003–2011/.

USAID (2024) DIV portfolio, https://divportal.usaid.gov/s/grant-portfolio.

Vance, A. (2016) *Elon Musk: How the Billionaire CEO of SpaceX and Tesla Is Shaping Our Future*, London: Virgin Books.

Varghese, C. and Kumar, S.S. (2022) Marginality: A critical review of the concept. *Review of Development and Change*, 27(1), 23–41.

Varoufakis, Y. (2023) *Technofeudalism: What Killed Capitalism*, New York: Random House.

Veblen, T. (1921) *The Engineers and the Price System*, Kitchener, ON: Batoche Books.

Vota, W. (2011) A great success: World Bank has 70% failure rate with ICT4D projects to increase Internet access, *ICTworks*, https://www.ictworks.org/great-success-world-bank-has-70-failure-rate-ict4d-projects-increase-universal-acces.

Vota, W. (2018) Fail Festival DC is dead. Long live failure, *ICTworks*, https://www.ictworks.org/fail-festival-dc-failure/.

Vota, W. (2024) Introducing the updated Principles for Digital Development, *ICTworks*, https://www.ictworks.org/updated-principles-for-digital-developent/.

Wagner, D.A., Day, B., James, T., Kozma, R.B., Miller, J., and Unwin, T. (2005) *Monitoring and Evaluation of ICT in Education Projects: A Handbook for Developing Countries*, Washington, DC: infoDev, World Bank.

Wahome, M.N.L. (2023) *Fabricating Silicon Savannah: The Making of a Digital Entrepreneurship Arena of Development*, Cham: Palgrave Macmillan.

Walker, D. (2024) 12 lessons from the weird, unnoticed end of Australia's bitter broadband war, *Club Troppo*, https://clubtroppo.com.au/2024/12/31/strange-death-of-australian-broadband-crisis.

Wallach, O. (2021) The world's tech giants, compared to the size of economies, *Visual Capitalist*, https://www.visualcapitalist.com/the-tech-giants-worth-compared-economies-countries/.

Waller, B.N. (2011) *Against Moral Responsibility*, Cambridge, MA: MIT Press.

Waller, B.N. (2014) *The Stubborn System of Moral Responsibility*, Cambridge, MA: MIT Press.

Walsh, A. (2024) Is the pursuit of money incompatible with morality? Some historical and philosophical reflections, in: Sandberg, J. and Warenski, L. (eds), *The Philosophy of Money and Finance*, Oxford: Oxford University Press, 171–190, https://academic.oup.com/book/55774/chapter-abstract/434287705.

Walsh, A. and Lynch, T. (2008) *The Morality of Money: An Exploration in Analytical Philosophy*, Basingstoke: Palgrave Macmillan.

Walsham, G. (2012) Are we making a better world with ICTs? Reflections on a future agenda for the IS field, *Journal of Information Technology*, 27(2), 87–93, https://journals.sagepub.com/doi/10.1057/jit.2012.4.

Walsham, G. (2017) ICT4D research: Reflections on history and future agenda, *Information Technology for Development*, 23(1), 18–41, https://www.tandfonline.com/doi/full/10.1080/02681102.2016.1246406.

Walzer, M. (1983) *Spheres of Justice*, New York: Basic Books.

Warner, J. and Ajibade, T. (2024) China's smart cities in Africa: Should the United States be concerned?, *Center for Strategic and International Studies*, https://www.csis.org/analysis/chinas-smart-cities-africa-should-united-states-be-concerned.

Waugaman, A. (2016) From principle to practice: Implementing the Principles for Digital Development, Washington, DC: The Principles for Digital Development Working Group https://www.slideshare.net/aszapla/implementing-the-principles-for-digital-development.

Webcertain (2024) *Webcertain Search and Social Report*, York: Webcertain, https://blog.webcertain.com/top-8-countries-for-time-spent-online-in-2024/11/07/2024/.

Wefers, S., Dieseth, T., George, E., Øverland, I., Jolapara, J., McAree, C., and Findlater, D. (2024) Understanding and deterring online child grooming: A qualitative study, *Sexual Offending: Theory, Research, and Prevention*, 19, e13147, https://sotrap.psychopen.eu/index.php/sotrap/article/view/13147.

Weigel, G. and Waldburger, D. (2004) ICT4D – *Connecting People for a Better World: Lessons, Innovation and Perspective of Information and Communication Technologies in Development*, Berne: Swiss Agency for Development and Cooperation and the Global Knowledge Partnership.

Weilan, Z. (2025) World's first humanoid robot boxing match kicks off in China, showcasing tech prowess, *Hangzhou China*, 6 May, https://www.ehangzhou.gov.cn/2025-05/26/c_293755.htm.

Weiler, R., Khan, A.W., Birger, R., and Schauer, T. (eds) (2005) *Information and Communication Technologies for Capacity-Building: Critical Success Factors*, Paris: UNESCO and Club of Rome, https://unesdoc.unesco.org/ark:/48223/pf0000189269.

Wells, T. (2011) Sen's capability approach, in: *Internet Encyclopedia of Philosophy*, https://iep.utm.edu/sen-cap/.

West, M. (2023) *An Ed-Tech Tragedy? Educational Technologies and School Closures in the Time of COVID-19*, Paris: UNESCO.

West, P. (2009) Open Educational Resources and Creative Commons Licenses, Commonwealth of Learning, 2009, https://oasis.col.org/server/api/core/bitstreams/0e5d31d1-4337-4462-916b-92e531fa7e61/content.

Wheeler, N.J. and Dunne, T. (2004) Moral Britannia? Evaluating the ethical dimension in Labour's foreign policy, London: Foreign Policy Centre, https://fpc.org.uk/publications/moral-britannia-evaluating-the-ethical-dimension-in-labours-foreign-policy.

Wheeler Fintech and Scale Team (2021) Scale and Fintech in Africa – some hard questions, Wheeler Institute for Business and Development, https://wheelerblog.london.edu/scale-and-fintech-in-africa-some-hard-questions/.

Whitbourne, S.K. (2012) Are you a neophiliac?, *Psychology Today*, https://www.psychologytoday.com/gb/blog/fulfillment-at-any-age/201203/are-you-a-neophiliac.

WHO (2016) *Monitoring and Evaluating Digital Health Interventions: A Practical Guide to Conducting Research and Assessment*, Geneva: WHO, https://www.who.int/publications/i/item/9789241511766.

WHO (2021) *Global Strategy on Digital Health 2020–2025*, Geneva: WHO, https://www.who.int/publications/i/item/9789240020924.

WHO (2024) Digital health, https://www.who.int/health-topics/digital-health.

Wikimedia Foundation (2022) New Wikipedia editor features make is easy for everyone to contribute, https://wikimediafoundation.org/news/2022/12/06/new-wikipedia-editor-features-make-it-easy-for-everyone-to-contribute/.

Williams, G. (2006) Responsibility, *Internet Encyclopedia of Philosophy*, https://iep.utm.edu/responsi/.

Williams, M. (2012) Agile projects: The people side, Government Digital Service Blog, https://gds.blog.gov.uk/2012/10/10/agile-projects-the-people-side/.

WIPO (2024) *Global Innovation Index 2024: Unlocking the Promise of Social Entrepreneurship*, Geneva: WIPO, https://www.wipo.int/web-publications/global-innovation-index-2024/assets/67729/2000-Global-Innovation-Index-2024_WEB3lite.pdf.

Wittes, B. and Chong, J. (2014) Our cyborg future: Law and policy implications, *Center for Technology Innovation at Brookings*, https://www.brookings.edu/wp-content/uploads/2016/07/cyborg_future_law_policy_implications_FINAL.pdf.

Wittgenstein, L. (2023) *Tractatus Logico-Philosophicus*, trans. Beaney, M., Oxford: Oxford University Press (first published in German in 1921).

Wood, E.M. (1990) The uses and abuses of 'Civil Society', *Socialist Register*, 26, 60–84, https://socialistregister.com/index.php/srv/article/view/5574.

World Bank (1981) IFC coins phrase 'emerging markets', World Bank Group Timeline, https://timeline.worldbank.org/en/timeline/eventdetail/2091.

World Bank (2002) *Information and Communication Technologies: A World Bank Group Strategy, April 2002*, Washington, DC: World Bank Group, http://documents.worldbank.org/curated/en/421471468325303463.

World Bank (2005) Universal Access & Output-Based Aid in Telecommunications and ICT, Washington, DC: GICT Policy Division, World Bank Group. https://documents1.worldbank.org/curated/en/106381468150319474/pdf/362330Universal0access01PUBLIC1.pdf.

World Bank (2022) The state of global learning poverty, 2022 update, https://thedocs.worldbank.org/en/doc/e52f55322528903b27f1b7e61238e416-0200022022/original/Learning-poverty-report-2022-06-21-final-V7-0-conferenceEdition.pdf.

World Bank (2024a) *Digital Progress and Trends Report 2023*, Washington, DC: International Bank for Reconstruction and Development/World Bank, https://www.worldbank.org/en/publication/digital-progress-and-trends-report.

World Bank (2024b) *Global Smart City Solutions Catalog 2024*, Washington, DC: World Bank, https://thedocs.worldbank.org/en/doc/46c2b2eea051820637bb269a521861b4–0090052024/original/Global-Smart-City-Solutions-Catalog–2024.pdf.

World Bank (2024c) Urban population growth (annual %), https://data.worldbank.org/indicator/SP.URB.GROW.

World Bank (2024d) Poverty and Inequality Platform, https://pip.worldbank.org/home.
World Bank (2024e) Digital progress and trends report: Interactive charts, https://www.worldbank.org/en/data/interactive/2024/03/04/digital-progress-and-trends-report-interactive-charts-digital-sector-development.
World Bank (n.d.) Poverty and inequality, https://datatopics.worldbank.org/world-development-indicators/themes/poverty-and-inequality.html.
World Economic Forum (2024) Mobilizing investment for clean energy in emerging economies, Cologny: World Economic Forum. https://initiatives.weforum.org/micee.
World Economic Forum and Capgemini (2024) The rise of global digital jobs, White Paper, 2024, Cologny: World Economic Forum, https://www3.weforum.org/docs/WEF_The_Rise_of_Global_Digital_Jobs_2024.pdf.
World Population Review (2025) Most technologically advanced countries, https://worldpopulationreview.com/country-rankings/most-technologically-advanced-countries.
WOUGNET (n.d.) About us, https://wougnet.org/about-us
WSIS (2003) Declaration of Principles. Building the Information Society: A global challenge in the new millennium, Geneva: ITU World Summit on the Information Society, https://www.itu.int/net/wsis/docs/geneva/official/dop.html.
Wynn-Williams (2025) *Careless People: A Cautionary Tale of Power, Greed, and Lost Idealism*, New York: Macmillan.
xfunnel (2025) What sources do AI search engines cite?, https://www.xfunnel.ai/blog/what-sources-do-ai-search-engines-choose.
Xi, J. (2014–2025) *The Governance of China*, Beijing: China International Publishing Group (5 vols).
XYZLab (2025) Full list of startup accelerators & incubators, https://www.xyzlab.com/startup-accelerators-incubators.
Yoon, S.Y., Lee, H.S., Zelt, T., Narlock, U., and Aguirre, E. (2020) *Smart City Pathways for Developing Asia: An Analytical Framework and Guidance*, Manila: ADB, https://www.adb.org/sites/default/files/publication/673441/sdwp-071-smart-city-pathways-developing-asia.pdf.
Yudkowsky, E. and Soares, N. (2025) *If Anyone Builds IT, Everyone Dies: Why Superhuman AI Would Kill Us All*, New York: Little, Brown and Company.
Zald, D.H., Cowan, R.L., Riccardi, P., Baldwin, R.M., Ansari, M.S., Li, R., Shelby, E.S., Smith, C.E., McHugo, M., and Kessler, R.M. (2008) Midbrain dopamine receptor availability use inversely associated with novelty-seeking traits in humans, *Journal of Neuroscience*, 28(53), 14372–14378, https://www.jneurosci.org/content/28/53/14372.
Zenger, J. and Folkman, J. (2013) The ideal praise to criticism ratio, *Harvard Business Review*, 15 March, https://hbr.org/2013/03/the-ideal-praise-to-criticism.
Zhenmin, L. (2019) Innovation tribute, https://www.uninnovation.network/innovation-tributes.
Zhou, Y., Zhing Lun Law Firm, with Practical Law China (2024) Regulation of telecommunications sector in China, https://uk.practicallaw.thomsonreuters.com/6-600-3886.
Zuboff, S. (2019) *The Age of Surveillance Capitalism: The Fight for a Human Future at the New Frontier of Power*, New York: PublicAffairs.
Zuiderwijk, A. and Janssen, M. (2014) The negative effects of open government data – investigating the dark side of open data, *dg.o '14: Proceedings of the 15th Annual International Conference on Digital Government Research*, 147–152, https://dl.acm.org/doi/10.1145/2612733.2612761.
Zwitter, A. and Boisse-Despiaux, M. (2018) Blockchain for humanitarian action and development, *Journal of International Humanitarian Action*, 3, 16, https://jhumanitarianaction.springeropen.com/articles/10.1186/s41018-018-0044-5.
Zwolinski, M. (ed.) (2014) *Arguing About Political Philosophy*, 2nd edition, London: Routledge.

Index